Photo Acoustic and Optical Coherence Tomography Imaging, Volume 2

Fundus imaging for the retina

Online at: https://doi.org/10.1088/978-0-7503-2056-6

Photo Acoustic and Optical Coherence Tomography Imaging, Volume 2

Fundus imaging for the retina

Edited by
Ayman El-Baz
University of Louisville, Louisville, KY, USA
and
University of Louisville at Alamein International University (UofL-AIU), New Alamein City, Egypt
Jasjit S Suri
AtheroPoint LLC, Roseville, CA, USA

IOP Publishing, Bristol, UK

ISBN 978-0-7503-2056-6 (ebook)
ISBN 978-0-7503-2054-2 (print)
ISBN 978-0-7503-2055-9 (myPrint)
ISBN 978-0-7503-2057-3 (mobi)

DOI 10.1088/978-0-7503-2056-6

Version: 20221201

IOP ebooks

British Library Cataloguing-in-Publication Data: A catalogue record for this book is available from the British Library.

Published by IOP Publishing, wholly owned by The Institute of Physics, London

IOP Publishing, No.2 The Distillery, Glassfields, Avon Street, Bristol, BS2 0GR, UK

US Office: IOP Publishing, Inc., 190 North Independence Mall West, Suite 601, Philadelphia, PA 19106, USA

With love and affection to my mother and father, whose loving spirit sustains me still.

—Ayman El-Baz

To my late loving parents, immediate family, and children.

—Jasjit S Suri

Contents

Preface

The book covers the state-of-the-art techniques of fundus imaging for the diagnosis of retinal diseases. Clinical disorders of the retina have been attracting the attention of researchers who are aiming to reduce the rate of blindness. These disorders include uveitis, diabetic retinopathy, macular edema, endophthalmitis, proliferative retinopathy, age-related macular degeneration (AMD), and glaucoma. Currently, most ophthalmologists perform diagnosis through visual observation and interpretation. Treatment is significantly dependent on having early and accurate diagnosis, which can be significantly improved by employing disease-specific computer-aided diagnostic (CAD) systems based on different imaging modalities, such as optical coherence tomography (OCT), fundus imaging, and optical coherence tomography angiography (OCTA). This book will focus on fundus imaging for the diagnosis of retinal diseases. Among the topics discussed in the book are texture interpretability of fundus imaging and diabetic retinopathy; a two-phase novel optic disc detection algorithm; enface image registration; automated detection and prediction of AMD; interobserver variability in the determination of diabetic retinopathy and quality of fundus images; Plus disease in preterm infants; retinal disease management using fundus autofluorescence images; mainstream ophthalmic AI algorithms: advances, limitations, and challenges; diabetic retinopathy detection and classification through fundus images using AlexNet; subtracting of the optic disk and the detection of exudates; and joint cup and disc segmentation.

In summary, the main aim of this book is to help advance scientific research within the broad field of fundus imaging for the diagnosis of retinal diseases. The book focuses on the major trends and challenges in this area, and it presents work aimed at identifying new techniques and their use in biomedical analysis.

Ayman El-Baz
Jasjit S Suri

Acknowledgements

The completion of this book could not have been possible without the participation and assistance of so many people whose names cannot all be listed. Their contributions are sincerely appreciated and gratefully acknowledged. However, the editors would like to in particular express their deep appreciation and indebtedness to Dr Ali H Mahmoud and Ahmed Naglah for their endless support.

Ayman El-Baz
Jasjit S Suri

Editor biographies

Ayman El-Baz

Ayman El-Baz is a Distinguished Professor at the University of Louisville, KY, USA, and the University of Louisville at Alamein International University (UofL-AIU), New Alamein City, Egypt. Dr El-Baz earned his BSc and MSc degrees in electrical engineering in 1997 and 2001, respectively. He earned his PhD in electrical engineering from the University of Louisville in 2006. Dr El-Baz was named as a Fellow of Coulter, AIMBE, and NAI for his contributions to the field of biomedical translational research. Dr El-Baz has almost two decades of hands-on experience in the fields of bio-imaging modeling and non-invasive computer-assisted diagnosis systems. He has authored or coauthored more than 700 technical articles (182 journal papers, 46 books, 97 book chapters, 253 refereed-conference papers, 214 abstracts, and 38 US patents and disclosures).

Jasjit S Suri

Jasjit S Suri is an innovator, scientist, visionary, industrialist, and an internationally known world leader in biomedical engineering. Dr Suri has spent over 25 years in the field of biomedical engineering/devices and management. He received his PhD from the University of Washington, Seattle, and his Business Management Sciences degree from Weatherhead, Case Western Reserve University, Cleveland, OH. Dr Suri was awarded the President's Gold medal in 1980 and was made a Fellow of the American Institute of Medical and Biological Engineering for his outstanding contributions. In 2018 he was awarded the Marquis Life Time Achievement Award for his outstanding contributions and dedication to medical imaging and its management.

List of contributors

Bejoy Abraham
College of Engineering Perumon, Kerala, India

Usman Akram
College of E&ME, NUST, Islamabad, Pakistan

Zahra Amini
Medical Image and Signal Processing Research Center, School of Advanced Technologies in Medicine, Isfahan University of Medical Sciences, Isfahan, Iran

Syed Anwar
Department of Software Engineering, UET Taxila, Pakistan

Sarforaz Bin Bashar
LV Prasad Eye Institute, Hyderabad, India

Jay Chhablani
University of Pittsburgh, Pittsburgh, PA, USA

Marco Dutra-Medeiros
Department of Ophthalmology, Central Lisbon Hospital Center, Lisbon, Portugal
and
NOVA Medical School, Universidade NOVA de Lisboa, Lisbon, Portugal
and
Portuguese Diabetes Association (APDP), Lisbon, Portugal
and
Portuguese Retina Institute, Lisbon, Portugal
and
Chronic Diseases Research Center (CEDOC), NOVA Medical School, Universidade NOVA de Lisboa, Lisbon, Portugal

Ayman El-Baz
Bioengineering Department, University of Louisville, Louisville, KY, USA

Wei Feng
Airdoc Co., Ltd, Beijing, China

Zongyuan Ge
Monash Medical AI Group, Monash University, Clayton, Australia

Mohammed Ghazal
Electrical, Computer and Biomedical Engineering, Abu Dhabi University, UAE

Abhilash Goud
LV Prasad Eye Institute, Hyderabad, India

Gowtham Chowdary Gudapati
IIT Hyderabad, India

Soumya Jana
IIT Hyderabad, India

Renu John
Indian Institute of Technology Hyderabad, Telangana, India

Lie Ju
Monash Medical AI Group, Monash University, Clayton, Australia

Rahele Kafieh
Medical Image and Signal Processing Research Center, School of Advanced Technologies in Medicine, Isfahan University of Medical Sciences, Isfahan, Iran

Farah Karim
Department of CS&IT, MUST, Pakistan

Samina Khalid
Department of CS&IT, MUST, Pakistan

Maleeha Khan
Department of Software Engineering, UET Taxila, Pakistan

Chandra Prakash Konkimalla
IIT Hyderabad, India

Ali Mahmoud
Bioengineering Department, University of Louisville, Louisville, KY, USA

Shanmukh Reddy Manne
IIT Hyderabad, India

Matilde Monteiro-Soares
Faculty of Medicine of the University of Porto, Portugal
and
CINTESIS@RISE, MEDCIDS, Faculty of Medicine of the University of Porto, Porto, Portugal
and
Portuguese Red Cross Health School Lisbon, Lisbon, Portugal

Ahmed Naglah
Bioengineering Department, University of Louisville, Louisville, KY, USA

Mehmet Nergiz
Department of Computer Engineering, Dicle University, Diyarbakır, Turkey

Zeinab Parsons
Department of Computer and Information Science, University of Michigan, Dearborn, USA

Ram Anudeep Peddoju
IIT Hyderabad, India

Niyas Punnakkadan
College of Engineering Trivandrum, Thiruvananthapuram, Kerala, India

Hossein Rabbani
Medical Image and Signal Processing Research Center, School of Advanced Technologies in Medicine, Isfahan University of Medical Sciences, Isfahan, Iran

Sivakumar Ramachandran
College of Engineering Trivandrum, Thiruvananthapuram, Kerala, India

Sílvia Rêgo
R&D Department, Fraunhofer Portugal AICOS, Porto, Portugal

Amnia Salma
Universitas Indonesia, Depok City, West Java, Indonesia

Tehmina Sheryar
Software Engineering, MUST, Pakistan

Filipe Soares
R&D Department, Fraunhofer Portugal AICOS, Porto, Portugal

Fatma Taher
Zayed University, College of Technological Innovation, Dubai, UAE

Mahnoosh Tajmirriahi
Medical Image and Signal Processing Research Center, School of Advanced Technologies in Medicine, Isfahan University of Medical Sciences, Isfahan, Iran

Anand Vinekar
Narayana Nethralaya Eye Hospital, Bangalore, India

Kiran Kumar Vupparaboina
University of Pittsburgh, Pittsburgh, PA, USA

Lin Wang
Harbin Engineering University, Harbin, China

Xin Wang
Airdoc Co., Ltd, Beijing, China

Yicheng Wu
Department of Data Science and AI, Faculty of Information Technology, Monash University, Clayton, Australia

Di Xu
Airdoc Co., Ltd, Beijing, China

Xin Zhao
Airdoc Co., Ltd, Beijing, China

IOP Publishing

Photo Acoustic and Optical Coherence Tomography Imaging, Volume 2

Fundus imaging for the retina

Ayman El-Baz and Jasjit S Suri

Chapter 1

Texture interpretability of fundus imaging and diabetic retinopathy: a review

Ahmed Naglah, Mohammed Ghazal, Fatma Taher, Ali Mahmoud and Ayman El-Baz

In this chapter a review of texture based methods for diagnosis and grading of diabetic retinopathy in fundus images is provided. We categorize these methods into major categories based on the style of model used and based on the approach taken to learn the features that correspond to the label. We provide a comparison between these methods in terms of classification accuracy. Following that, we perform an exploratory study to investigate the possibility of using a deep learning guided approach to produce image gradient filters. We include a qualitative analysis of fundus texture using deep learning guided texture versus traditional hand-crafted texture features.

1.1 Introduction

Diabetic retinopathy (DR) is a highly prevalent complication of diabetes that features dilation in the blood vessels (microaneurysms) and then develops to the growth of new blood vessels (neovascularization) in the retina. The proliferation stage is associated with vascularization activity leading to damage to the blood vessels and eventually blindness [1, 2]. The worldwide population with diabetes mellitus (DM) is 463 million in 2019 as per the International Diabetes Federation (IDF) with a very high projected trend through the upcoming decades [3]. Blindness caused by diabetic retinopathy had an estimated prevalence of 18.5% in 2020 [4]. The increase in impairment and blindness as DR complications in the year 2015 compared to 1990 is suggested to be caused by the growing prevalence of type 2 diabetes in countries with a low income and middle income countries in particular [5].

We can categorized DR into four levels, starting from mild nonproliferative diabetic retinopathy (NPDR), moderate NPDR, severe NPDR, and reaching the end stage of proliferative diabetic retinopathy (PDR), which can be associated with

doi:10.1088/978-0-7503-2056-6ch1

severe complications [6–8]. The early detection and diagnosis of DR can prevent the damage and reduce the risk of complications. Some markers can indicate the evolution of DR in diabetic patients. Among the early signs of DR, healthcare professionals can consider findings from optical coherence tomography angiography, blood vessel oxygen saturation, and other circulatory/vascular markers [9].

Continual advances in the field of medical imaging opened the door and enriched the opportunities to discover new biomarkers for DR. These biomarkers can be detected by clinical examination or can be automatically detected using artificial intelligence methods. Healthcare professionals can have better insight on possible biomarkers using the advanced technologies that enable retinal imaging, such as spectral-domain and swept source-based optical coherence tomography (OCT), fundus autofluorescence (FAF), OCT angiography, and ultrawide field imaging and FFA for the early detection of DR, as discussed in the study by Markan *et al* [10]. Some of these markers are suggested to have links to diabetic macular edema which is a common complication of DR itself in addition to vision loss [10].

In this chapter, we will review the recent advances in using texture based methods for fundus images towards developing DR classifications models. First, we will describe selected studies that use texture for DR level classification. We will categorize these methods into major categories based on the style of model used and based on the approach taken to learn the features that correspond to the label. Following that, we will explore an open dataset for DR (APTOS 2019 [11]) to describe selected aspects for which we can perform texture based analysis. Finally, we will perform an exploratory study to investigate the possibility of using a deep learning guided approach to produce image gradient filters. We will explore the results obtained by these filters qualitatively on the APTOS 2019 dataset [11]. Figure 1.1 shows selected samples from the APTOS 2019 dataset [11].

1.2 DR interpretability approaches

One of the main disadvantages of convolutional neural networks (CNNs) is the difficulty of performing interpretations on the images because the standard feature maps will eventually lead to the merging of spatial parts in the images, in particular at the edge of the fully connected stage. There are multiple techniques in the literature that have been explored recently that investigate the possibility of addressing this issue. Each technique addresses the problem from a certain perspective and has certain limitations. Also some of these techniques are based on hand-crafted features or brute-force approaches. In the following subsections we discuss the main broad categories of approaches that can be used to enhance the interpretability of DR. We will discuss the strengths and weaknesses of each category and we will discuss why we choose our proposed approach.

1.2.1 Attention based models

The first category of models for interpretability is attention based networks. Attention models contain an attention layer(s) at which a recurrent neural network is used spatially to determine the spatial activation patterns.

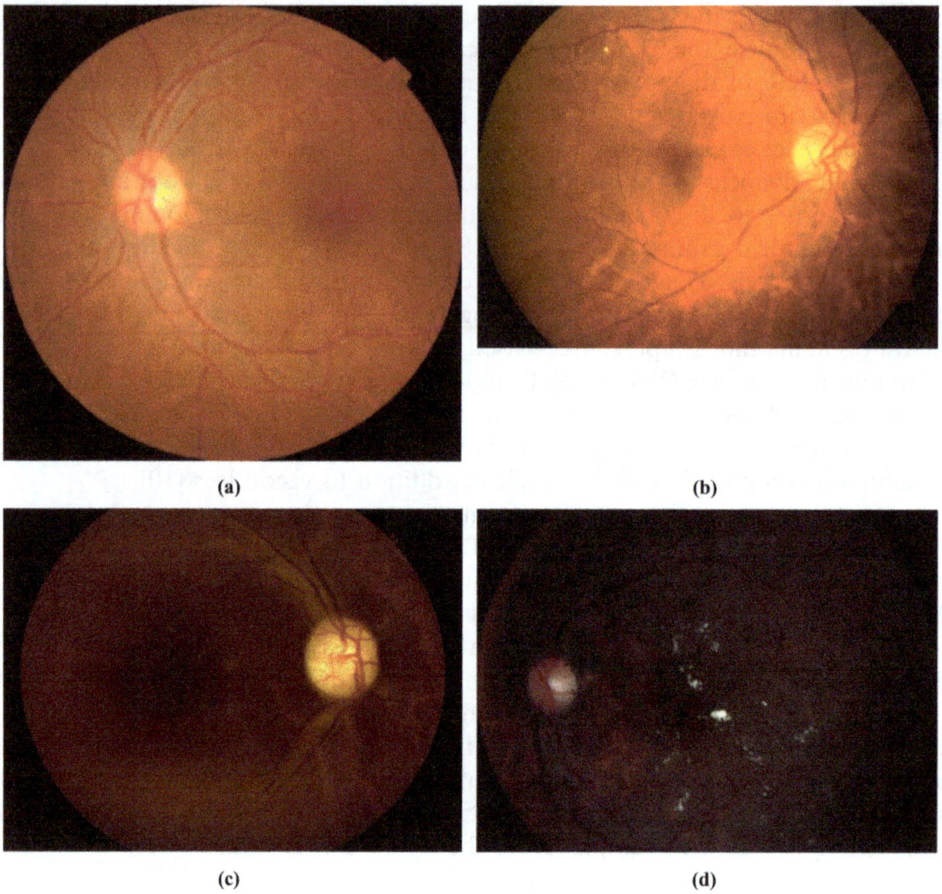

Figure 1.1. Samples from the APTOS 2019 [11] dataset which contains fundus images of positive ((a) and (c)) and negative ((b) and (d)) diabetic retinopathy (DR) cases, and it also includes the four stages of DR. One of the challenges in these images is the color and shade variabilities as can clearly seen between (a) versus (c) and (b) versus (d).

A study by He *et al* studied defined region-wise features for each DR grade to enhance the discrimination capability [12]. Moreover, through the model itself, a global attention block (GAB) was deployed to help in obtaining localized feature maps for the highly spread fundus images [12]. A cross disease attention network was deployed in [13] to utilize the localized maps from another deep learning network specified for diabetic macular edema. This is a good alternative to applying the concept of transfer learning, similar to [14].

The attention concept was also deployed in a different way to enable fusing multi-view images of fundus images. Precise diagnosis and grading of DR each require multi-view fundus imaging with around four or more photographs for the retina. The fusion of these views cannot be combined without addressing the spatial correspondence between the detected features from each view. Also if one view is used, the corresponding information about the lesion from the other views will not be considered in the classification task. The study by Luo *et al* proposed an attention based fusion network in order to facilitate this fusion or multi-instance learning problem [15]. Closely related attention concepts were used in different studies aiming towards having interpretable variables within the model itself in order to offer this to the users [16–18].

1.2.2 Feed-forward style localization maps

The second category of interpretability approaches involves having spatial maps that are updated during the feed-forward and back-propagation phases without having any attention layers or blocks. These models usually use domain specific maps that are taken into consideration in the loss function during the back-propagation stage.

Specific examples of this category are illustrated in the study by Gondal *et al* [19], in which a class activation maps layer is placed just before the output layer to replace the dense layers. That layer is constructed of a pooling performed on a number of rectangular windows (*K* windows) at the latest feature maps. The output from this pooling is fully connected to each output pin that corresponds to a certain class. Analyzing the weights of that layer can help in generating heatmaps for each class which can consequently represent the regions that correspond to each DR grade.

Regression activation maps (RAMs) can be used as one of the approaches to localize the regions that correspond to a maximum contribution for the output. Wang *et al* presented a RAM based method to provide a visual explanation for DR [20]. The way the architecture of the RAM is built ensures having information for prediction at the output, in addition to having a map between the last convolutional feature map and input images, which can help in localizing the associated area with the regression outcome [20]. DR is also explored through testing with concept activation vectors (TCAV) [21]. This technique uses so called concept activation vectors (CAVs), which are designated vectors at the latent space that define a certain concept (which can be interpreted in the data domain). This vector is estimated and then can be tested on the class level using the TCAV score. This scope represents the sensitivity across entire classes of inputs [21]. The strength of this study is that the score can be used to detect the class at which the diagnosis can be sensitive to error. Given a DR problem with five classes, their results shows that the deep learning model sometimes inaccurately predicts DR level 1 as DR level 2 [21]. Another study by Lin *et al* addressed the interpretability in DR [22]. They proposed a framework they named explainable lesion learning and generation (ELLG) for the interpretability of DR automatic detection methods and to achieve more accurate diagnosis, based on a gated multi-layer saliency map (GMSM), inspired by class activation mapping [22].

1.2.3 Occlusion based models

Given the different anatomical regions of interest in the retina, occlusion (removal from the input image) of these specific regions one at a time, then monitoring the quantitative metrics of DR detection can help in localizing the regions of high importance. There are not many studies that address this in the literature of DR.

It worth mentioning that the intended approach in this subsection is not related to retinal vein occlusion as a medical condition [23–28], despite being related to DR and to the fundus images. The intended occlusion is a spatial removal of a whole region of interest (ROI) in the input image while conducting the deep learning prediction. The removal can be performed at a pre-processing stage using methods such as those in [29], or by manual segmentation. In the study by Grassmann *et al* the same concept is applied to fundus images to study age-related macular degeneration, and the results show a change in the detection accuracy within the occluded areas [30].

1.2.4 Hand-crafted texture analysis methods

There are several studies that address the problem of DR using hand-crafted texture analysis methods. A study by Shafi *et al* used Kirsch's convolution kernels to extract the map of blood vessels in the retina, then multiple-order statistical texture features were used to extract features for use by statistical classifiers [31]. This approach shows promising quantitative results with an accuracy of 95.19% compared to other techniques including those involving CNNs. In this study the vessels were extracted from the images before the classification stage. However, the whole image field of view was used in the classification, which contrasts it to the occlusion based methods discussed earlier.

Another study by Rehman *et al* used a discrete wavelet transform for the histogram equalized images in order to prepare the features for the classifier [32]. Certain features were selected from the wavelet coefficients which show the sparse nature of feature selection within this work. In a study by Jadhav *et al* a segmentation stage was added before extracting features [33]. That segmentation would enable the detection of microaneurysms in fundus images within the classification model.

1.2.5 Other approaches

Other studies have handled the problem of DR AI interpretability through multiple approaches. Some of these studies used the uncertainty of labels or inter-observer variability [34, 35]. Other studies took this further to the generative AI domain [36]. Other approaches and techniques are used as well [12, 37–39]. Multiple studies handled the interpretability by studying the classification of fundus images at each color channel separately [40–43].

Machine learning algorithms can be deployed to predict certain clinical biomarkers such as insulin and duration of diabetes, which can indicate high risk populations for diabetic retinopathy [44]. From the deep learning side, visual maps would be an important option for interpretation by an ophthalmologist. Such maps can be generated by altered neural networks such as that proposed by De La Torre *et al* [45]. Merging interpretable variables (such as clinical markers or multiple instance learning MIL from the images) and fused deep learning can be a good alternative in terms of having a middle point between hand-crafted features and automated features. A similar technique was deployed in the study by Costa *et al* [46]. We believe the accuracy of the deep learning model for DR detection is not the main target, because in practice, users, physicians, and healthcare professionals want to understand the insights and the interpretation behind DR detection. These interpretations are usually taken forward for further investigation in the pre-clinical domain and *in vitro* studies.

The study conducted by Mohsin *et al* [47] shows the use of each color channel separately in order to classify DR in images. These channels are the red, green, and blue from the RGB domain, which are uncommonly used for analyzing color images.

1.3 Texture challenges in a fundus imaging dataset

In this review and exploratory study, we use the APTOS 2019 [7] dataset. The dataset contains 5590 fundus images with five levels. Level 0 represents no DR with no noticeable abnormalities in the retina. Level 1 represents mild DR with only microaneurysms which is a localized disturbance in the diameters of blood vessels in the retina. Level 2 represents moderate microaneurysms that do not qualify as nonproliferative diabetic retinopathy. Level 3 represents severe DR with no signs of proliferating DR. Level 4 represents proliferating DR that features neovasculariza-tion and or haemorrhage.

1.3.1 Intensity based analysis is challenging for DR classification

The images in this dataset have varying dimensions and some of them have a varying field of view. This can lead to uncertainty in observing the prediction outputs. For that reason, approaches that are built on occlusion or that try to analyze certain textures might be useful for having better interpretations in this case. Also given that there are a lot of overlaps between the classes in the gray scale and all HSV channels (see figure 1.2), threshold based methods have a higher probability of failure in obtaining a discriminant edge between classes.

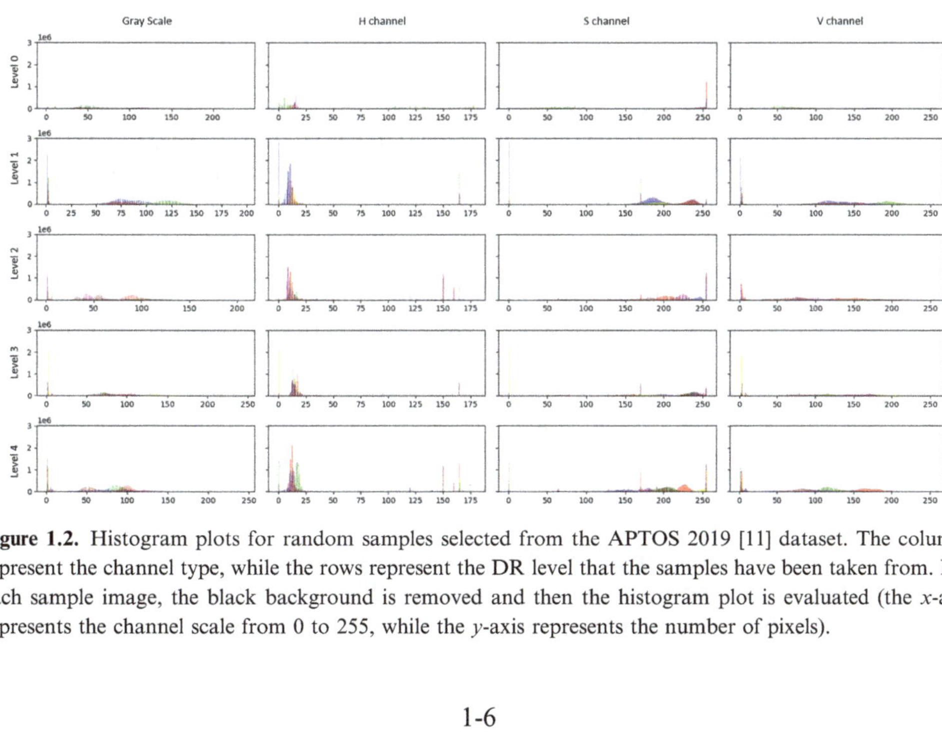

Figure 1.2. Histogram plots for random samples selected from the APTOS 2019 [11] dataset. The columns represent the channel type, while the rows represent the DR level that the samples have been taken from. For each sample image, the black background is removed and then the histogram plot is evaluated (the x-axis represents the channel scale from 0 to 255, while the y-axis represents the number of pixels).

1.3.2 Image gradient analysis is challenging for DR classification

By going to higher orders, image gradients can be included in the texture computations in order to capture the possible variabilities across images. Image gradients are the basis of traditional machine learning models that are based on hand-crafted features, or that are based on feature descriptors, such as ORB (oriented FAST and rotated BRIEF) [48], SIFT [49], and SURF [50]. In our explorative study, we will use basic image gradient filters in order to visualize the image gradients between DR classes. In our analysis, we used a Laplacian filter [51] and Sobel filter [52] to capture the first and second order gradients in various directions.

Based on our qualitative analysis, figure 1.3 shows that there might be possible discriminant borders between DR levels based on the texture values. However, as we can see in figures 1.4–1.8, there is possible intra-class variability in each DR level, which can act in the opposite direction as a challenge to achieve accurate classification between classes.

Moreover, the anatomical features within the retina can contribute to these image gradient maps in a difficult-to-measure way. Accordingly, gradient analysis can be further combined with occlusion methodologies and/or segmentation methods in order to separate the factors coming from these anatomical features.

In the study by Ismail *et al* fractional-order edge detection masks were used to diagnose DR [53]. Having a fractional order in the hand-crafted filter can add a degree of freedom to the model which can be optimized to achieve the best results. In their work, they proposed a fractional-order Sobel filter and their qualitative results shows that the generated maps can be used to classify between normal, early stage, and late stage DR [53].

Combining multiple factors can help in achieving a better diagnosis. In the study by Sayres *et al* [54], heatmaps for deep learning generated predictions in addition to image gradients were presented along with the fundus images to a panel of ophthalmologists and retina specialists. The results show that the grading accuracy is improved for readers with model assistance than without for image grading assistance only. Image gradients plus heatmaps have partial improvement (improvement in patients with DR but lower accuracy for patients without DR).

1.4 Detecting texture patterns with the aid of deep learning

As we observed in the previous section, hand-crafted features can show an indication of the DR level in qualitative examination. However, there remains the persistent challenge of how to select these filters in order to develop a model that can achieve generalizability. There is a certain level of generalizability that is required for the model to be implemented in the real world and be reproduced on independent datasets. Most of the hand-crafted feature based models do not qualify for this level of generalizability because usually the selection of filters, such as Sobel filters, Laplacians, or even fractional-order filters [53], requires tuning for the hyper-parameters to obtain the fit on a certain dataset. This hyper-parameter tuning can be shifted dramatically when changing the scanner or the image acquisition device.

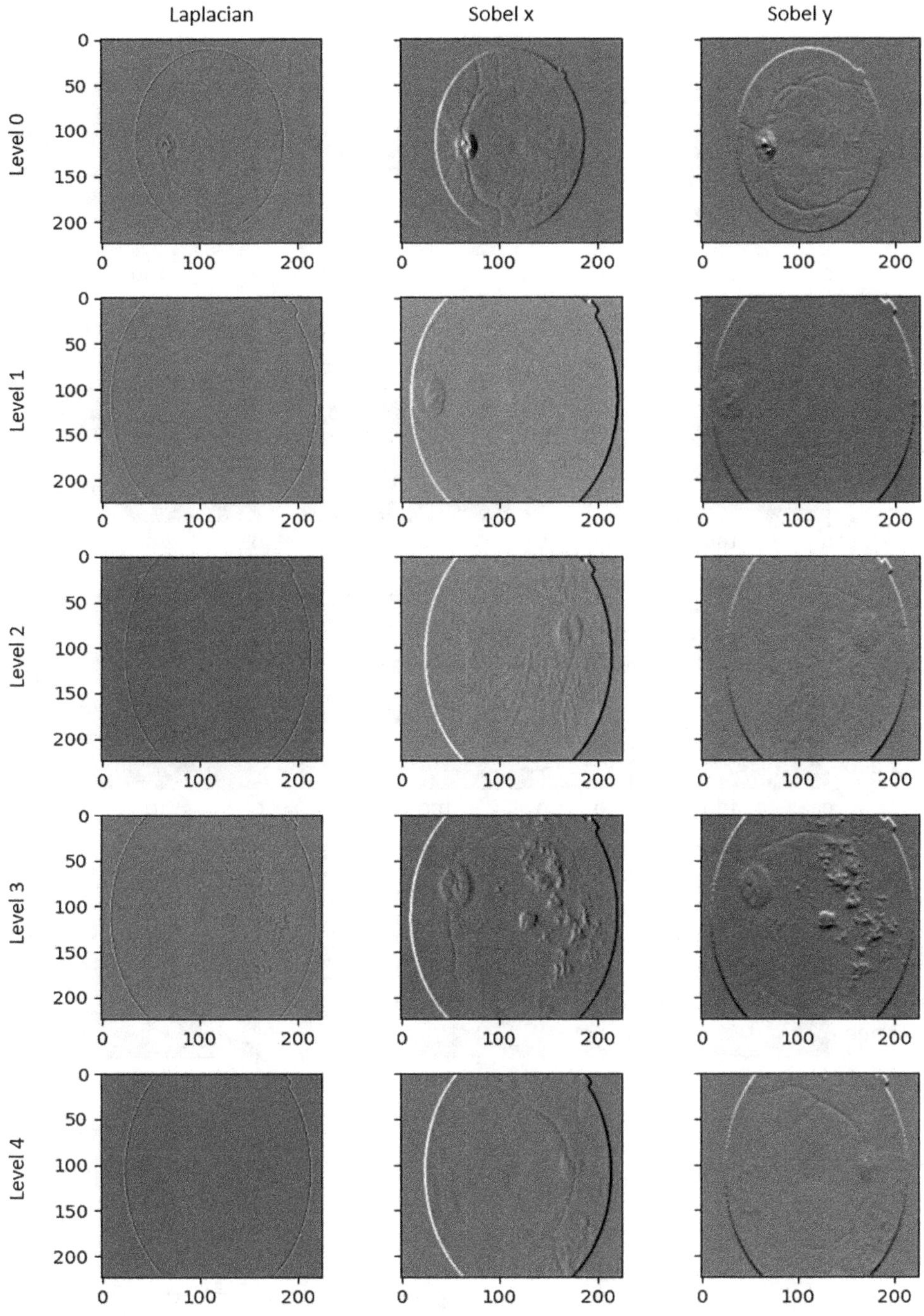

Figure 1.3. Visualization of image gradient features for random samples selected from each DR. For the samples shown, we can see that a discriminant border might exist between levels. However, this visual appearance might be misleading once further analysis is performed to explore the intra-class variability between samples.

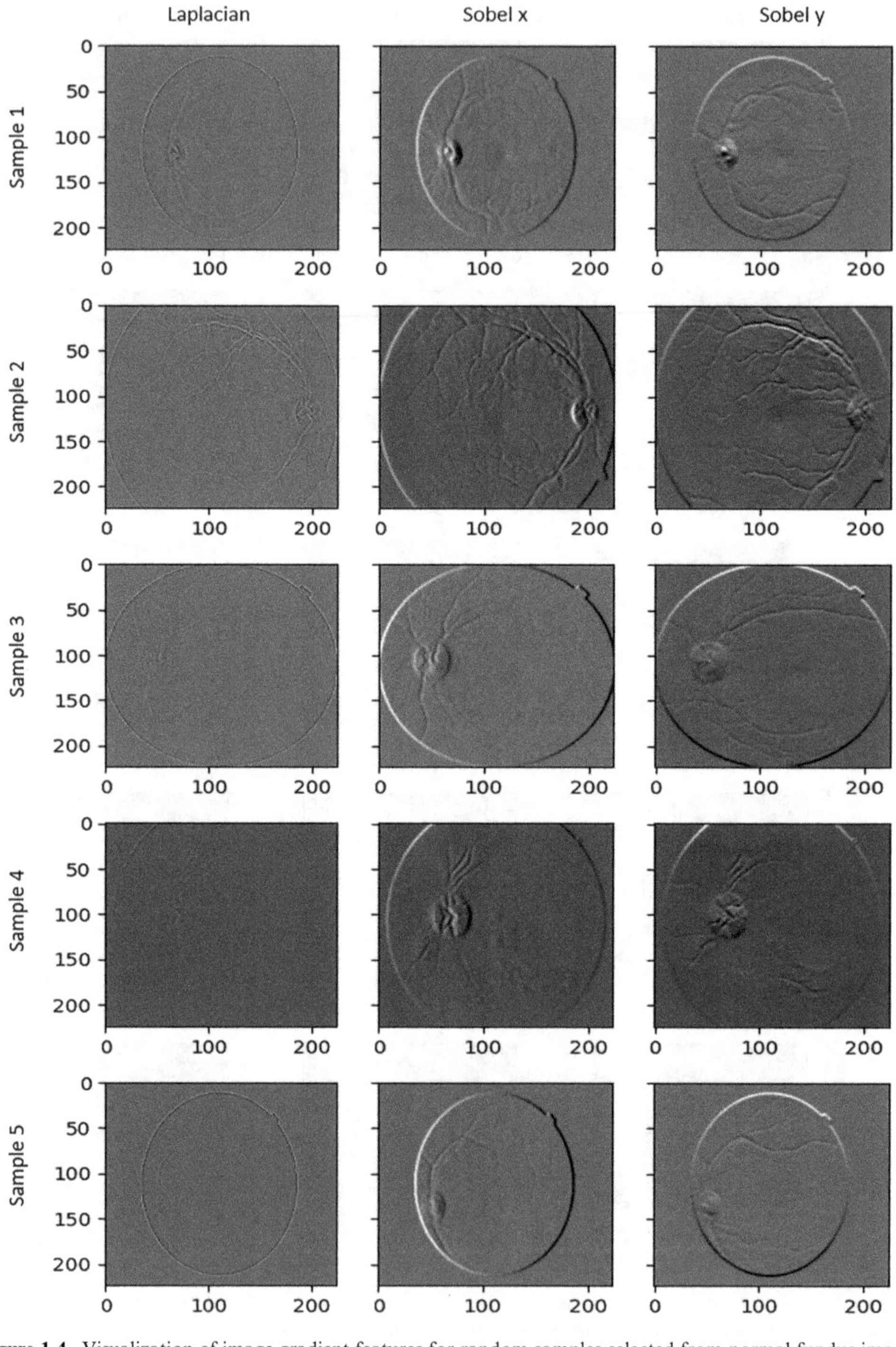

Figure 1.4. Visualization of image gradient features for random samples selected from normal fundus images (DR level 0). For the samples shown, we can see a level of intra-class heterogeneity which can be a challenge for DR classification tasks.

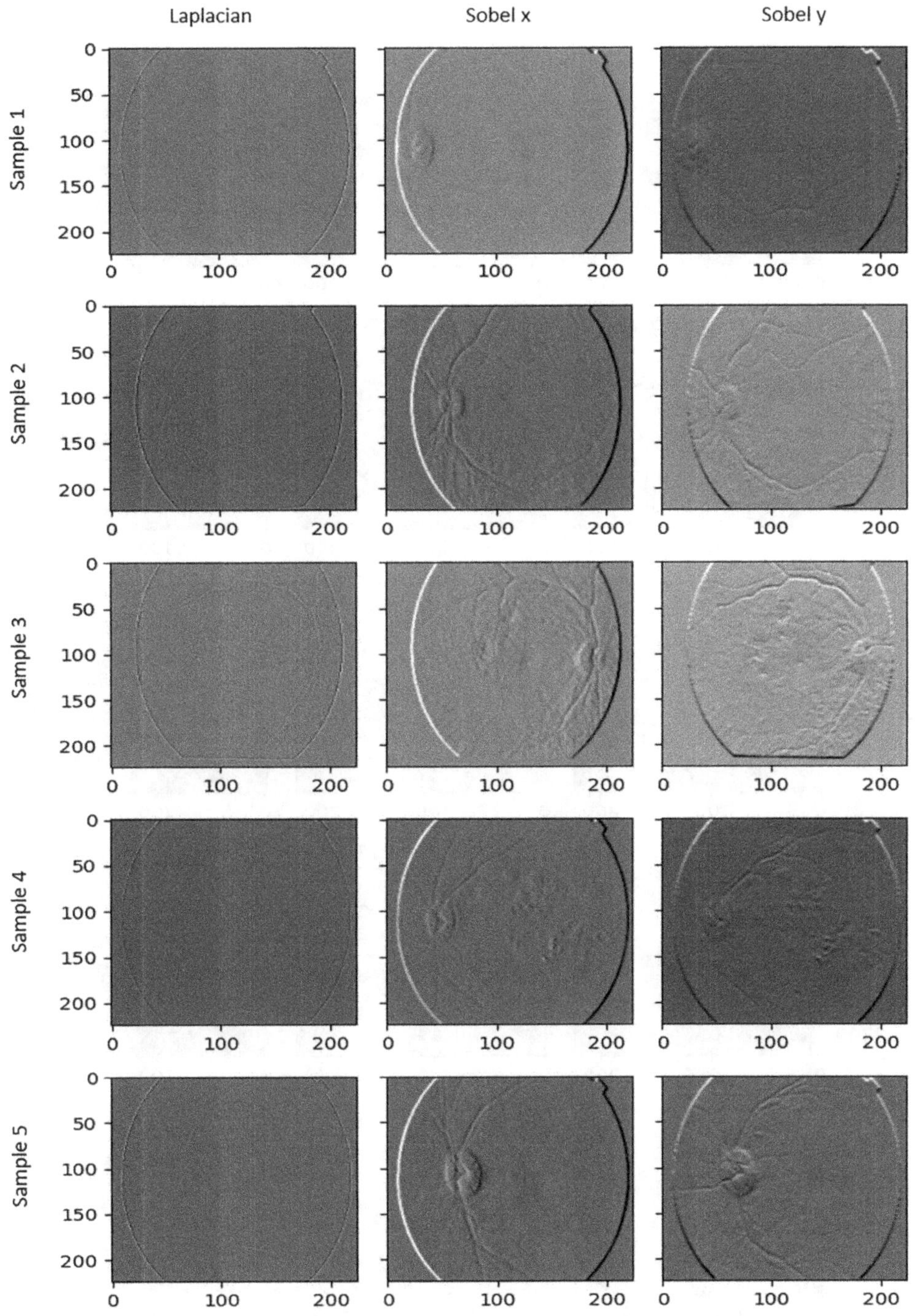

Figure 1.5. Visualization of image gradient features for random samples selected from fundus images of DR level 1. For the samples shown, we can see a level of intra-class heterogeneity which can be a challenge for DR classification tasks.

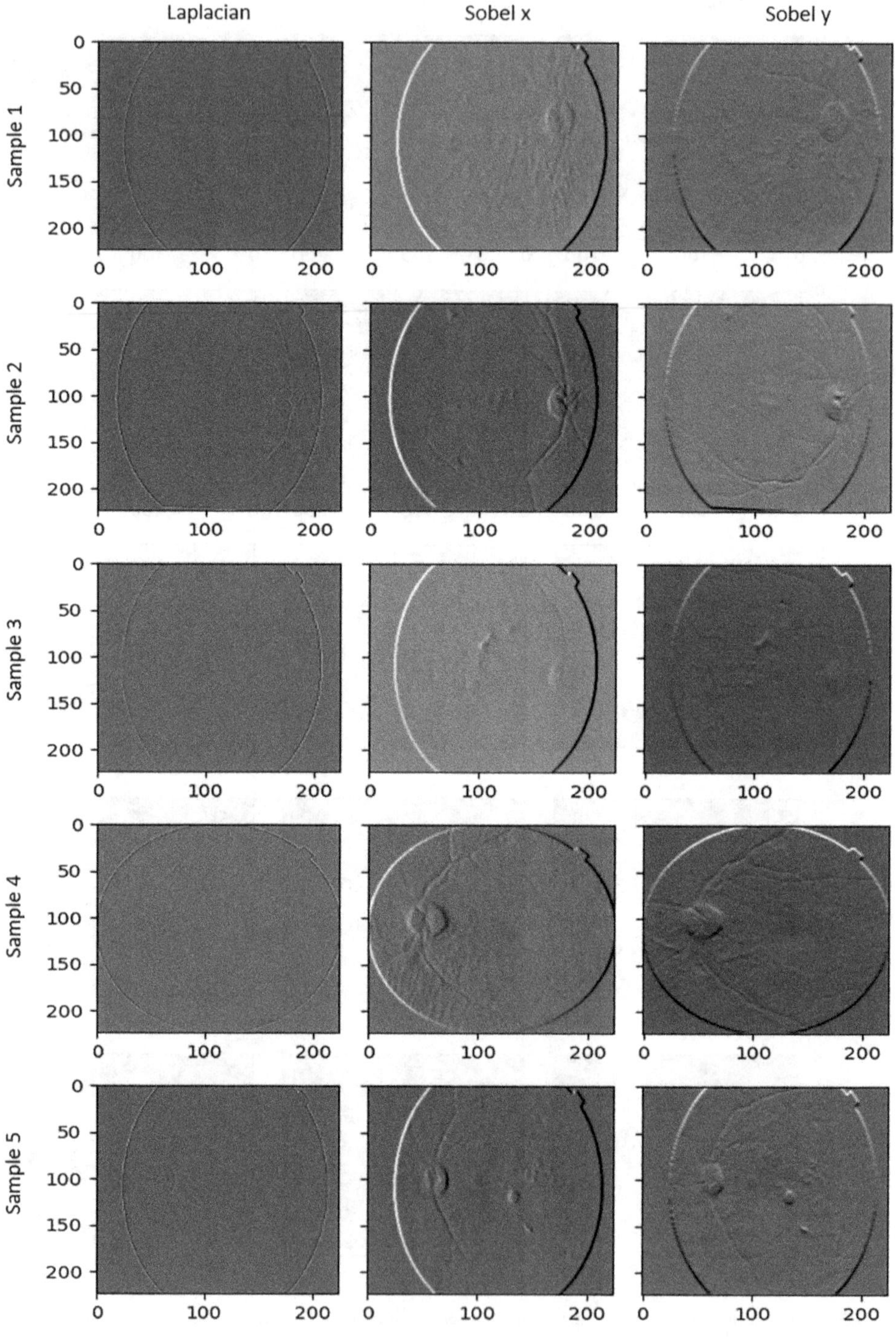

Figure 1.6. Visualization of image gradient features for random samples selected from fundus images of DR level 2. For the samples shown, we can see a level of intra-class heterogeneity which can be a challenge for DR classification tasks.

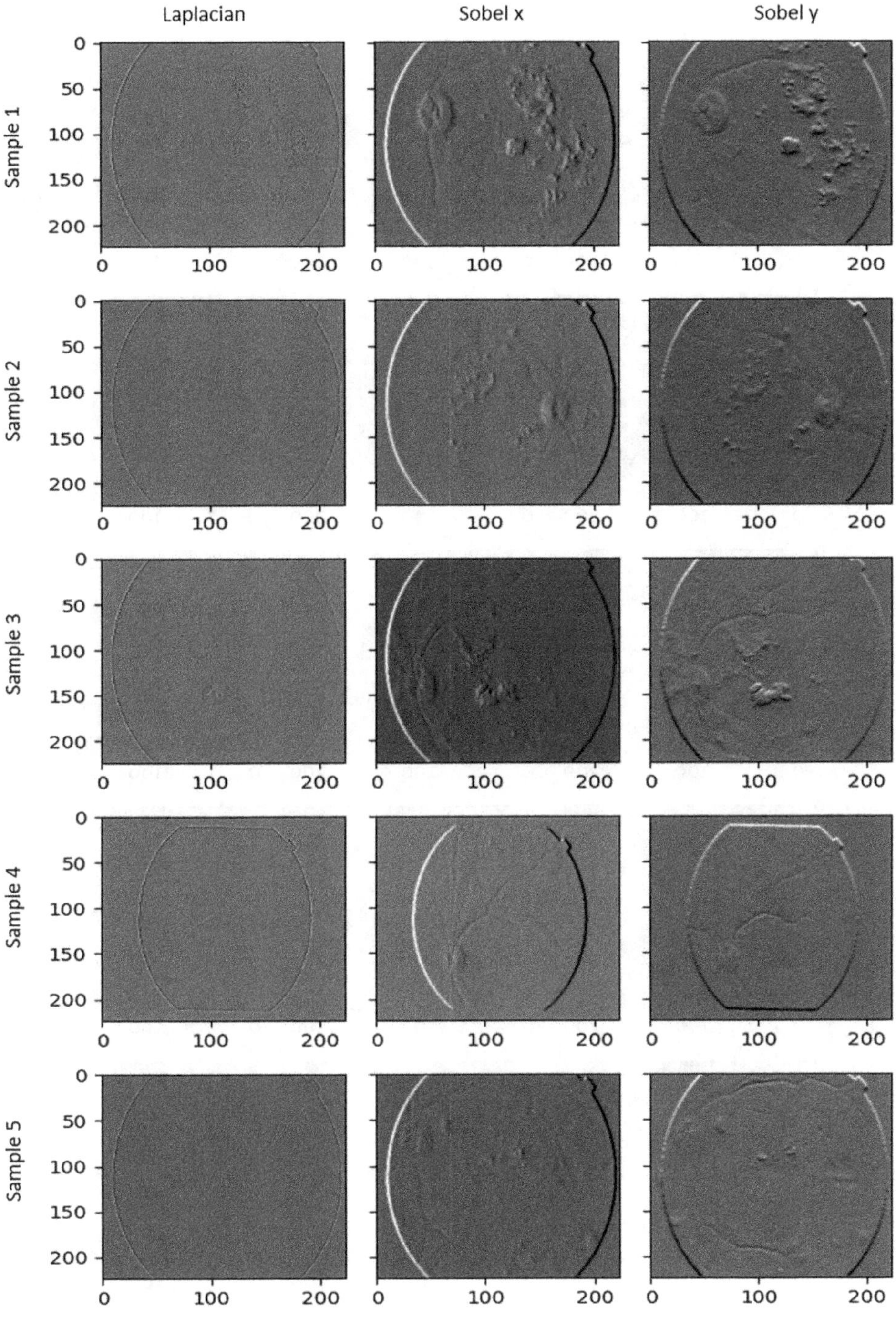

Figure 1.7. Visualization of image gradient features for random samples selected from fundus images of DR level 3. For the samples shown, we can see a level of intra-class heterogeneity which can be a challenge for DR classification tasks.

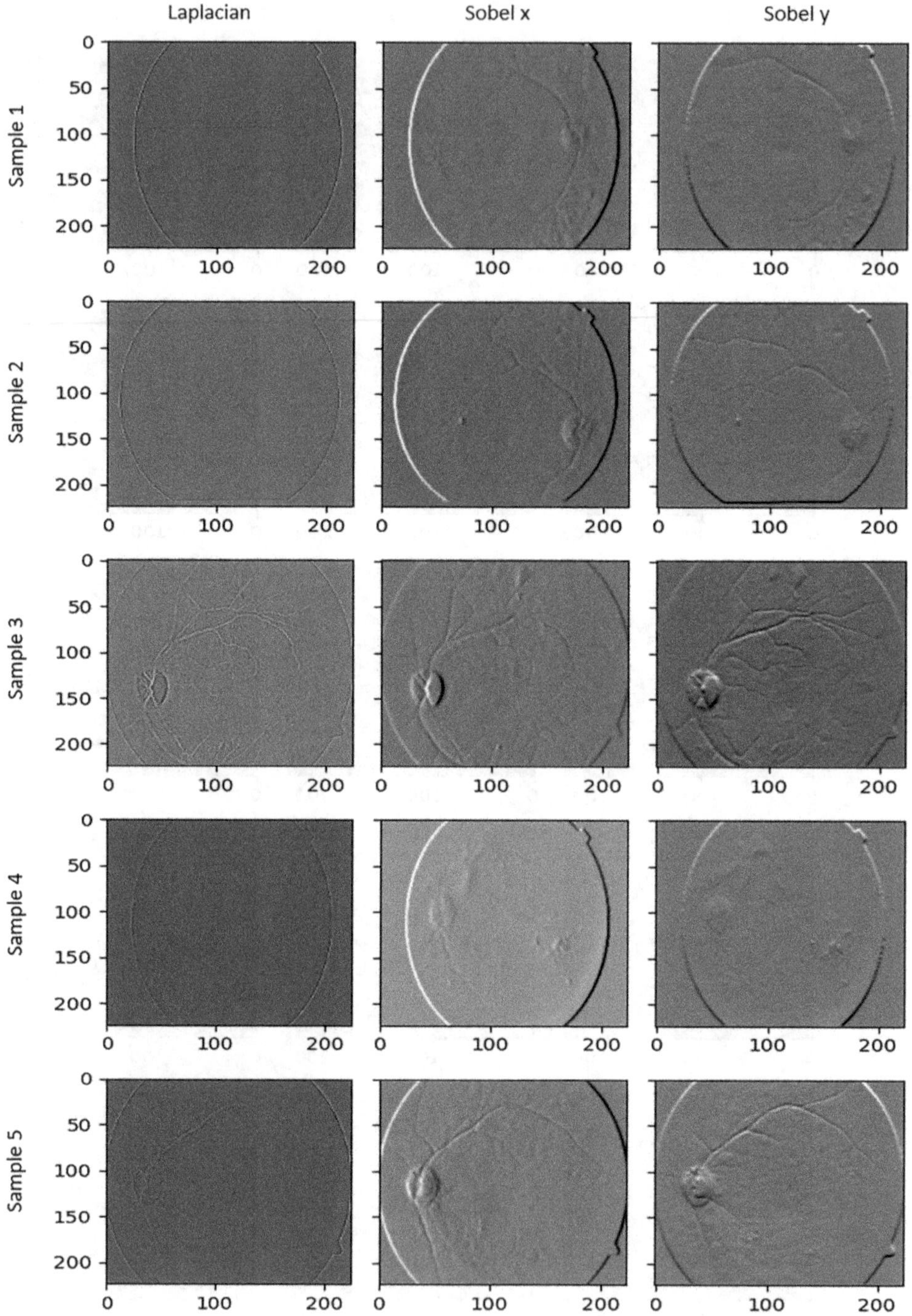

Figure 1.8. Visualization of image gradient features for random samples selected from fundus images of DR level 4. For the samples shown, we can see a level of intra-class heterogeneity which can be a challenge for DR classification tasks.

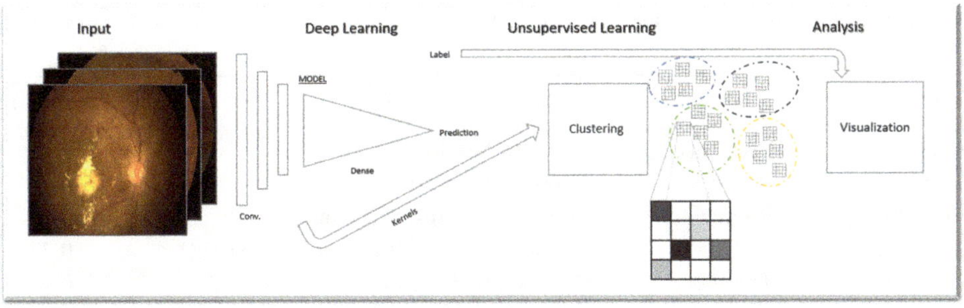

Figure 1.9. Illustrative diagram that present the proposed kernel analysis pipeline within this exploratory study. Texture interpretability can be performed at different levels. One of these levels is in the kernel domain of the filtered after finalizing the training phase of the deep neural network. The kernels can be extracted from the network and analyzed in order to estimate the dominant kernels towards the prediction outcome. One of the possible applications of the resulting kernels is to deploy a fast algorithm that can perform quick screening for DR using the pre-obtained filters.

For that reason, in this study we will explore a deep learning based approach that might be used to automatically select the suitable filters that can be used for feature detection and therefore for accurate classification of DR. In this approach, inspired by [55], we will train a CNN deep learning model to classify DR in fundus images. After finishing the training process, we will extract the filters from the CNN layers, and we will perform clustering on these kernels in order to obtain the dominant pattern in these filters. Finally we will visualize the image gradients generated by these filters to assess the possibility of having a discriminant pattern between the DR levels. The proposed pipeline is illustrated in figure 1.9.

1.4.1 Convolutional neural network model

The first stage of the proposed pipeline is a CNN based deep learning model. There are recent advances in the field of CNN for classification, such as [56–58]. However, the concept behind our approach is to use a more simple CNN model, in order to capture distinctive filters that can be interpreted in the later stages. That can be inspired by weak classifier based models in which each classifier is weak, but the collaboration of all classifiers can produce improved results.

In this exploratory work, the choice for our CNN was the VGG-16 model [59] as one of the early adopted CNNs for classification. We performed transfer learning from the ImageNet weights [60], then we proceeded with training the model on our dataset. After finalizing the training phase, we extracted the kernels from the first layer, and we used these kernels for the second stage of our pipeline.

For the sake of reviewing the classification accuracy of fundus images using different models, we trained the baseline VGG-16 on the APTOS 2019 [11] dataset starting from random weights to estimate the baseline of our classification problem (see table 1.1). In the same table we also make a comparison with the quantitative results on fundus images on different datasets and models. For the sake of kernel extraction, we performed transfer learning from ImageNet [60] to initialize the

Table 1.1. A review of the DR classification quantitative results that involve texture based features including dataset, number of classes, the model used, and accuracy.

Dataset	Classes	Model	Accuracy
APTOS 2019 [11]	Five classes	VGG-16 baseline	0.49
APTOS 2019 [11]	Binary	VGG-16 baseline	0.51
APTOS 2019 [11]	Binary	VGG-16 + transfer learning	0.55
APTOS 2019 [11]	Binary	VGG-16 transfer learning + SVM [14]	0.77
IDRiD [32]	Three classes	Hand-crafted texture features [31]	0.95
MESSIDOR [61]	Binary	DWT + SVM [62]	0.97
MESSIDOR [61]	Binary	DWT + KNN [62]	0.98
IDRiD [32]	Binary	Segmentation + texture features + SVM [33]	0.81
EyePACS [63]	Five classes	CNN + gray scale [47]	0.97
EyePACS [63]	Five classes	CNN + red [47]	0.94
EyePACS [63]	Five classes	CNN + green [47]	0.94
EyePACS [63]	Five classes	CNN + blue [47]	0.97

weights of all layers except the last layer, which was randomly initialized, then we trained the network on the referenced dataset to estimate the relevant clusters.

1.4.2 Kernel clustering

After finalizing the training process on the DR dataset, we applied a clustering method in an unsupervised way to discover the patterns in the filters/kernels produced by the CNN. We use agglomerative hierarchical clustering [64] with $k = 3$ to facilitate visualizations. Since we have observed that the training of the CNN can be considered a set of weak classifiers, the unsupervised learning conducted by the clustering method served as a boosting method to estimate hand-crafted filters guided by a deep learning process. For visualization purposes, we min–max normalized the output of the filtered images to be in the range of 0–255 on the gray scale.

1.4.3 Visualization

After extracting and processing the kernels, we applied filtering for the sake of visualizing the texture patterns in fundus images similar to what was performed in figures 1.3–1.8 but with the deep learning guided kernels.

Based on our qualitative analysis, figure 1.10 shows that there might be possible discriminant borders between the DR level based on the texture values. In future work, these kernels can be used to feed a machine learning method to predict DR grades. In this work, a comparison to the standard image gradient filters illustrated in figure 1.3 can show the effect of deep learning guidance of the quantitative results. However, as we can see in figures 1.11–1.15, there is possible intra-class variability in each DR level, which can act in the opposite direction as a challenge to achieve accurate classification between classes.

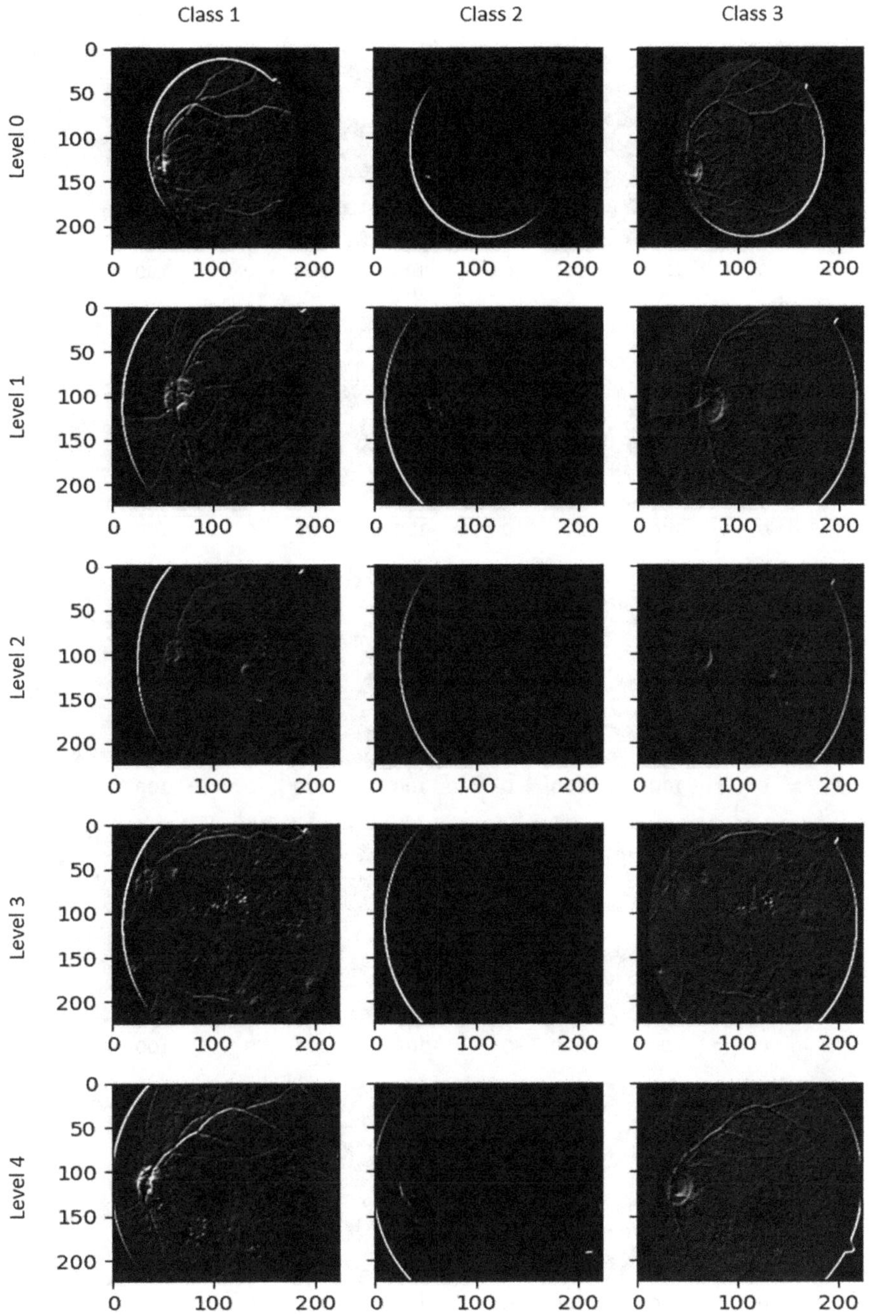

Figure 1.10. Visualization of image gradient features for random samples selected from each DR using the resulting kernels produced by the clustering analysis of the trained CNN network. For the samples shown, we can see that a discriminant border might exist between levels. However, this visual appearance might be misleading once further analysis is performed to explore intra-class variability between samples.

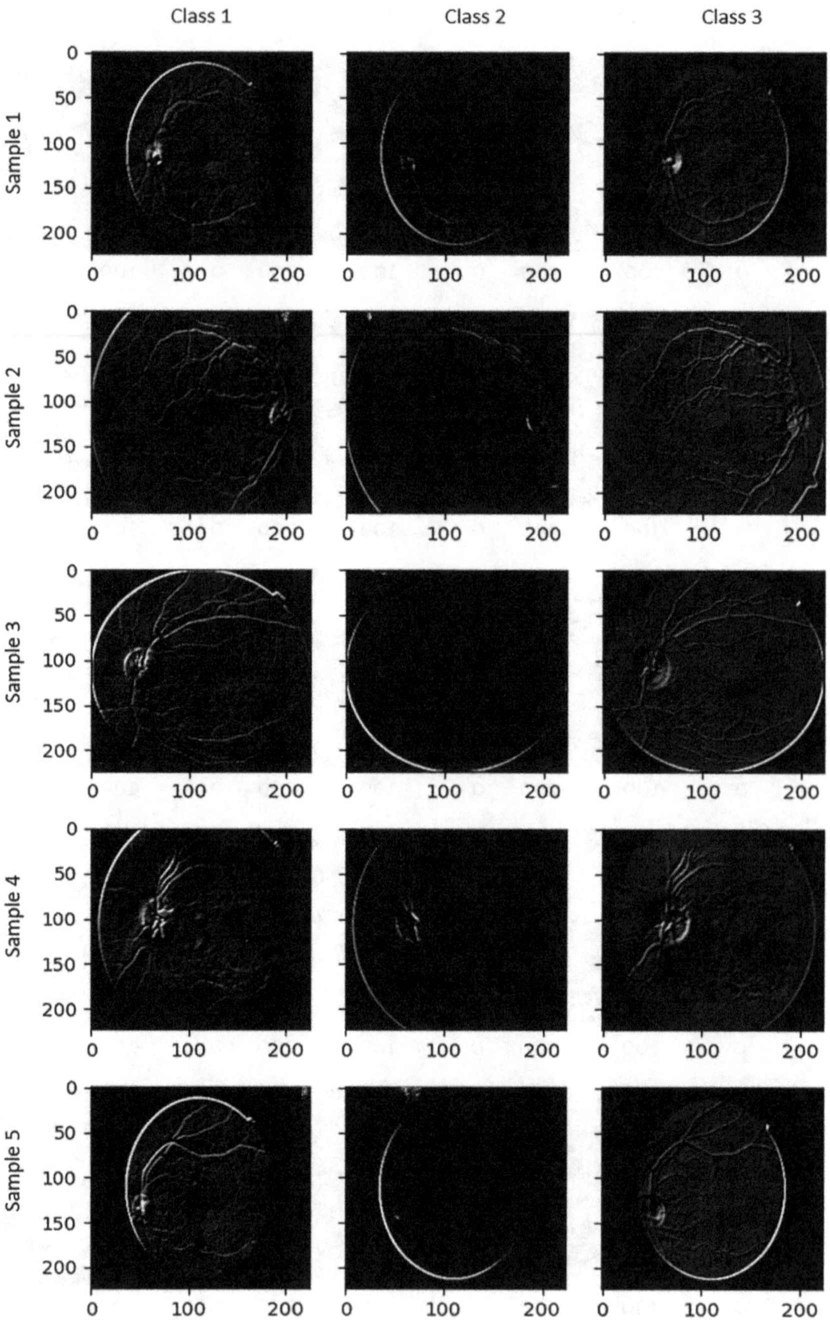

Figure 1.11. Visualization of image gradient features for random samples selected from normal fundus images (DR level 0) using the resulting kernels produced by the clustering analysis of the trained CNN network. For the samples shown, we can see a level of intra-class heterogeneity which can be a challenge for DR classification tasks.

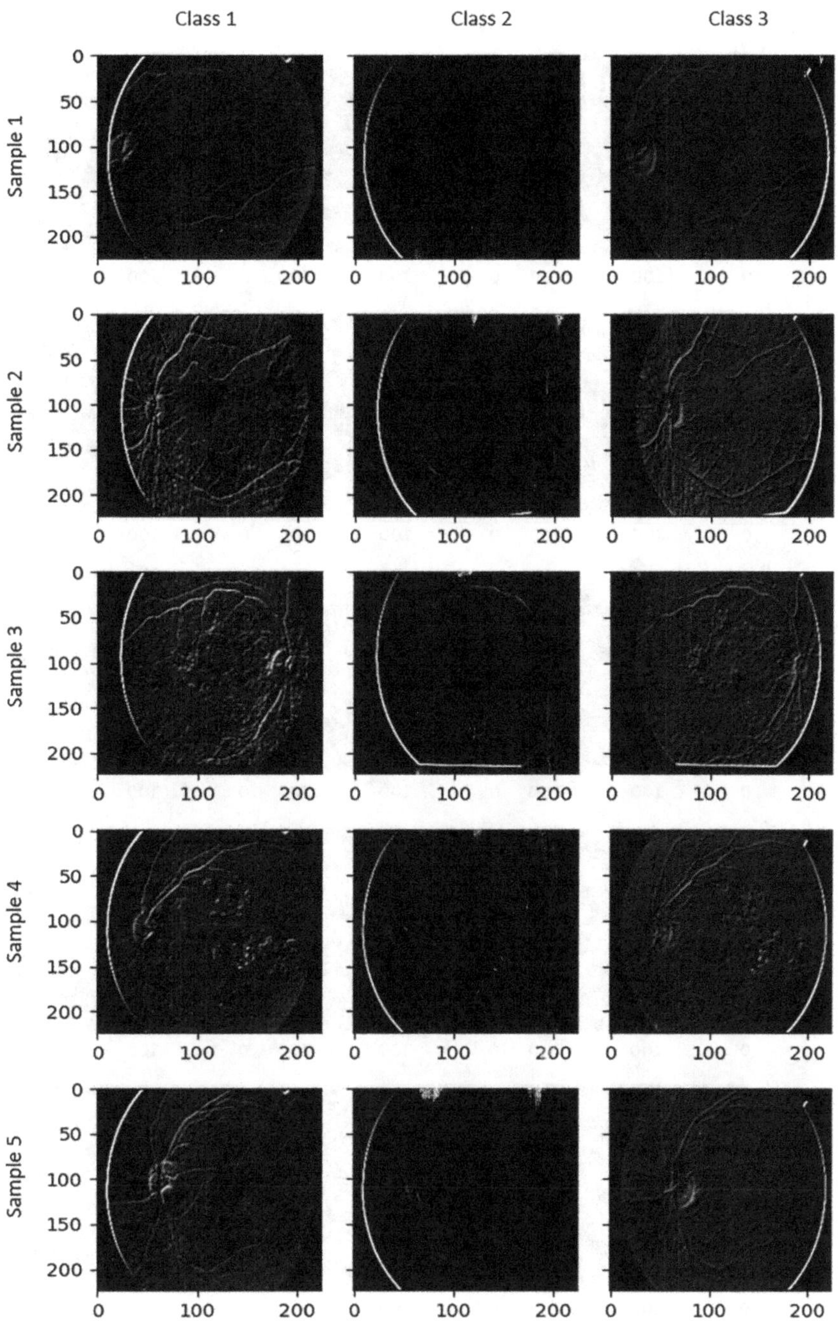

Figure 1.12. Visualization of image gradient features for random samples selected from fundus images of DR level 1 using the resulting kernels produced by the clustering analysis of the trained CNN network. For the samples shown, we can see a level of intra-class heterogeneity which can be a challenge for DR classification tasks.

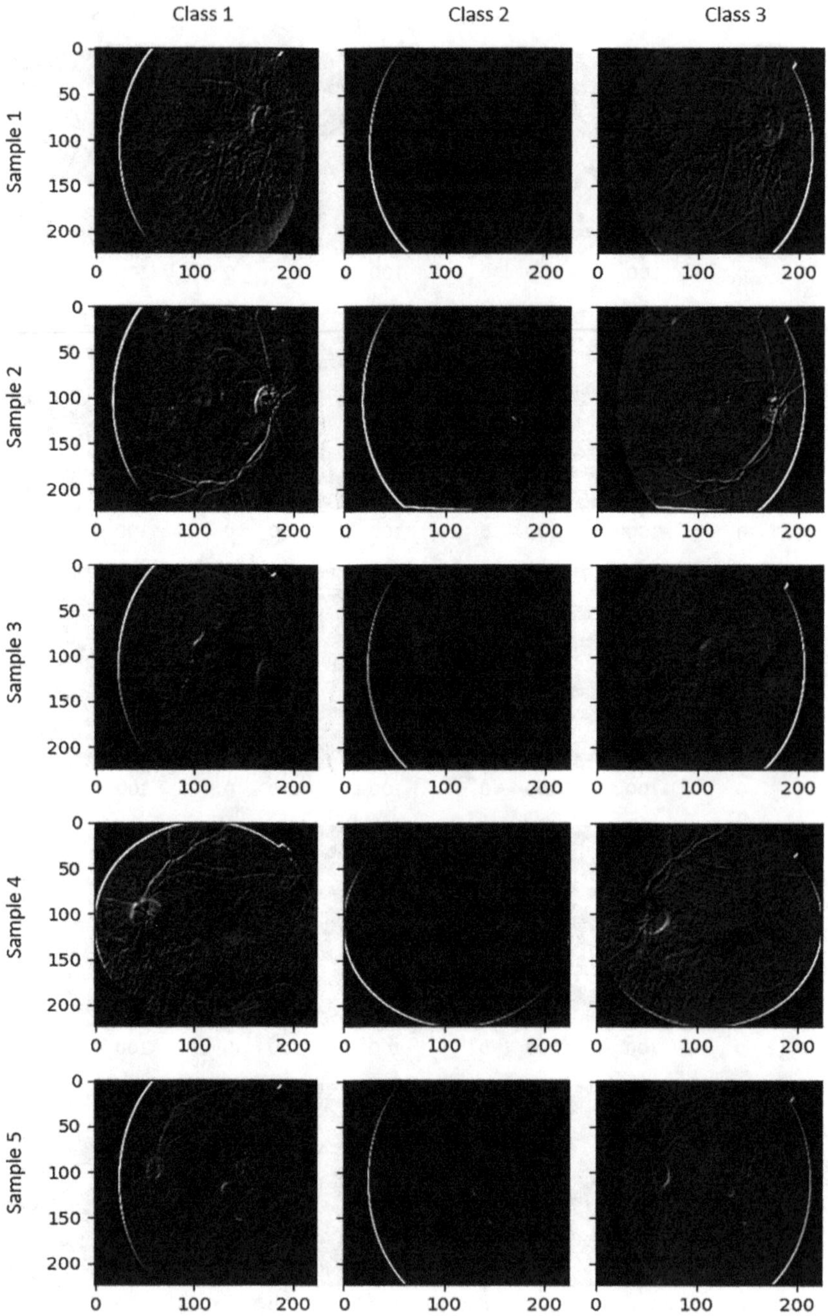

Figure 1.13. Visualization of image gradient features for random samples selected from fundus images of DR level 2 using the resulting kernels produced by the clustering analysis of the trained CNN network. For the samples shown, we can see a level of intra-class heterogeneity which can be a challenge for DR classification tasks.

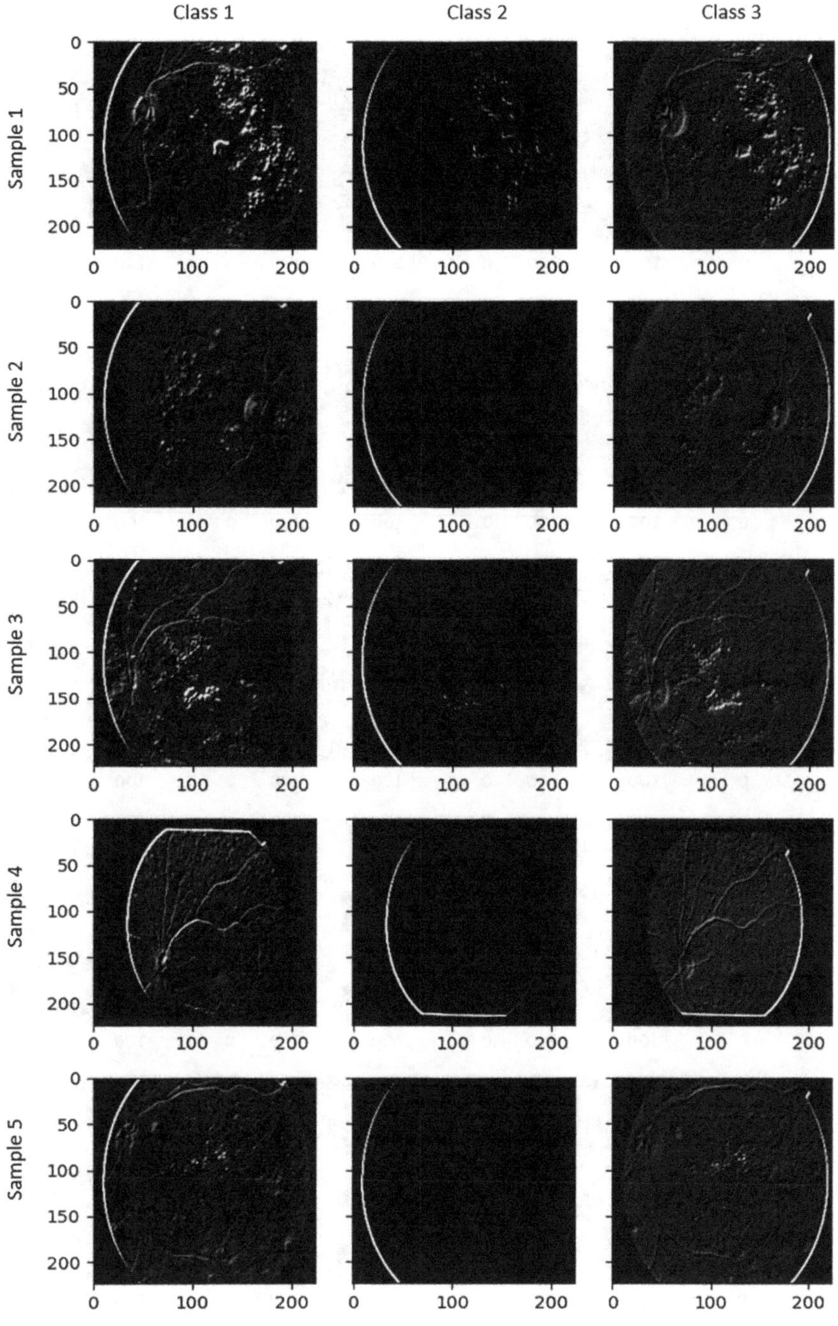

Figure 1.14. Visualization of image gradient features for random samples selected from fundus images of DR level 3 using the resulting kernels produced by the clustering analysis of the trained CNN network. For the samples shown, we can see a level of intra-class heterogeneity which can be a challenge for DR classification tasks.

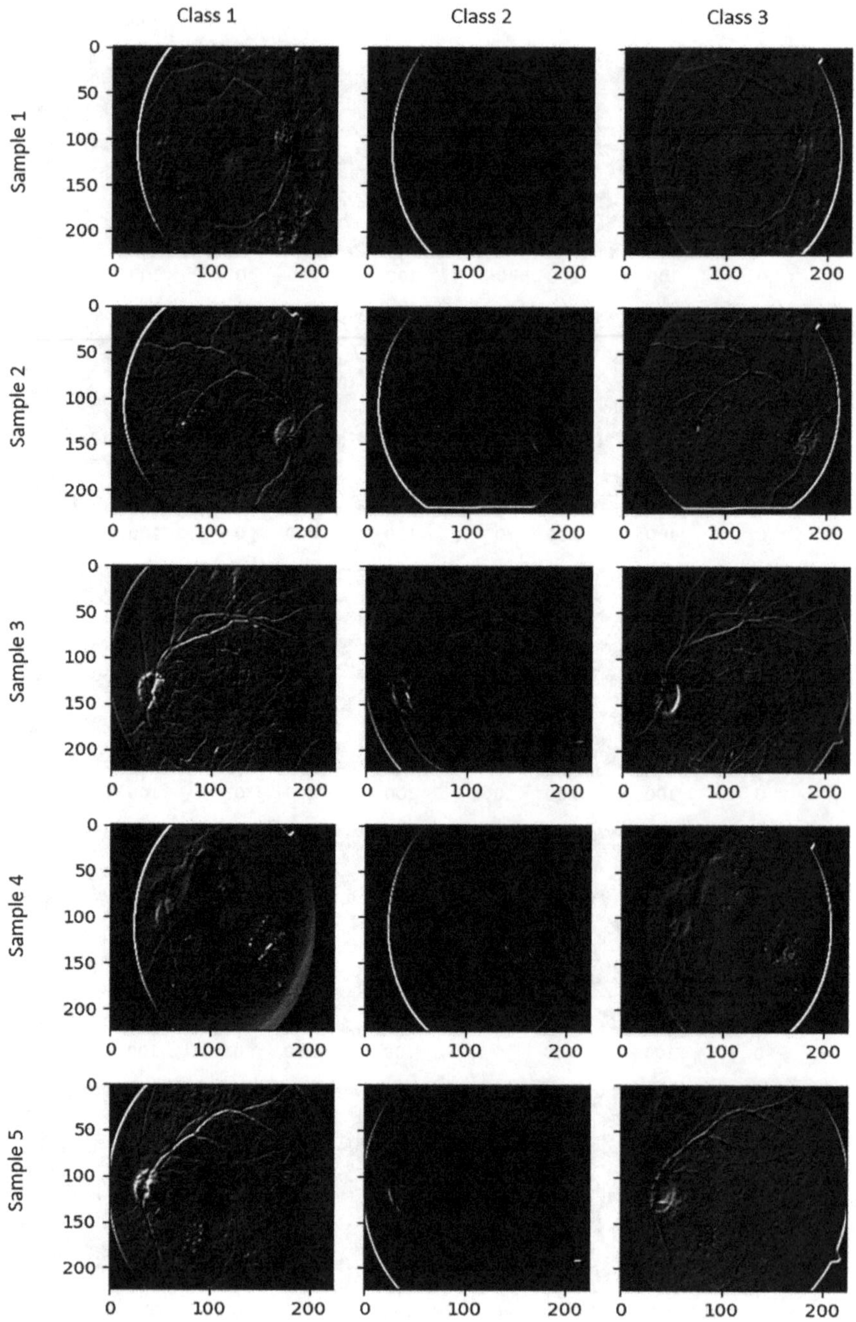

Figure 1.15. Visualization of image gradient features for random samples selected from fundus images of DR level 4 using the resulting kernels produced by the clustering analysis of the trained CNN network. For the samples shown, we can see a level of intra-class heterogeneity which can be a challenge for DR classification tasks.

One of the possible approaches that can be taken forward to address this is to combine occlusion based methods and texture features. Only parts of the images can be highly associated with DR. Accordingly, detection for certain anatomical regions of interest or removal of irrelevant regions from the fundus images are required. Also, one of the possible options is to design fractional-order edge detection masks similar to [53], after taking the weights produced by the deep learning model. Having pre-defined weights can be a benefit in particular for real-time systems that cannot afford the processing costs of deep learning inference.

1.5 Limitations and challenges

Since fundus images contain large areas that might not be relevant to the anatomical regions of interest for DR, the produced kernels from this study can be subject to noise generated by these regions. To enhance the performance of the proposed model, an object detection algorithm needs to be incorporated to detect the regions of interest to be boosted or removed from the system.

AI learning models are subject to generalizability and reproducibility limitations. The proposed method was applied on the referenced fundus imaging dataset. Certain adjustments and re-training needed to be performed before applying them to other datasets. The applicability of texture based methods can be highly sensitive to the area on which the texture is applied. Accordingly, the selection of a relevant object detection or area localization algorithm can be of high importance to this topic.

1.6 Conclusion

In this chapter we reviewed the recent texture based advances and methods for fundus images in the problem of DR classification. We went through selected studies that used texture for DR detection and grading. We categorized these methods into major classes based on the style of model used and based on the approach taken to learn the corresponding texture features. Following that, we explored an open dataset and we performed an exploratory study to investigate the possibility of using a deep learning guided approach to produce image gradient filters. Qualitative results show that deep learning can be used to suggest image gradient features that can be taken further into the problem of DR grading.

In addition to the retina [65–74], this work could also be applied to various other applications in medical imaging, such as the prostate [75–79], kidney [80–98], heart [99–116], lungs [117–166], brain [167–189], vascular system [190–200], bladder [201–205], liver [206, 207], head and neck [208, 209], and injury prediction [210] as well as several non-medical applications [211–217].

References

[1] Stitt A W *et al* 2016 The progress in understanding and treatment of diabetic retinopathy *Prog. Retin. Eye Res.* **51** 156–86

[2] Engerman R L 1989 Pathogenesis of diabetic retinopathy *Diabetes* **38** 1203–6

[3] Teo Z L *et al* 2021 Global prevalence of diabetic retinopathy and projection of burden through 2045: systematic review and meta-analysis *Ophthalmology* **128** 1580–91

[4] Steinmetz J D *et al* 2021 Causes of blindness and vision impairment in 2020 and trends over 30 years, and prevalence of avoidable blindness in relation to VISION 2020: the right to sight: an analysis for the global burden of disease study *Lancet Glob. Heal.* **9** e144–60

[5] Vujosevic S, Aldington S J, Silva P, Hernández C, Scanlon P, Peto T and Simó R 2020 Screening for diabetic retinopathy: new perspectives and challenges *Lancet Diab. Endocrinol.* **8** 337–47

[6] Wang W and Lo A C Y 2018 Diabetic retinopathy: pathophysiology and treatments *Int. J. Mol. Sci.* **19** 1816

[7] Kusuhara S, Fukushima Y, Ogura S, Inoue N and Uemura A 2018 Pathophysiology of diabetic retinopathy: the old and the new *Diab. Metab. J.* **42** 364–76

[8] Honasoge A, Nudleman E, Smith M and Rajagopal R 2019 Emerging insights and interventions for diabetic retinopathy *Curr. Diab. Rep.* **19** 100

[9] Safi H, Safi S, Hafezi-Moghadam A and Ahmadieh H 2018 Early detection of diabetic retinopathy *Surv. Ophthalmol.* **63** 601–8

[10] Markan A, Agarwal A, Arora A, Bazgain K, Rana V and Gupta V 2020 Novel imaging biomarkers in diabetic retinopathy and diabetic macular edema *Ther. Adv. Ophthalmol.* **12** 507–24

[11] Karthick M and Sohier D 2019 APTOS 2019 Blindness Detection. Kaggle https://kaggle.com/competitions/aptos2019-blindness-detection

[12] He A, Li T, Li N, Wang K and Fu H 2020 CABNet: category attention block for imbalanced diabetic retinopathy grading *IEEE Trans. Med. Imaging* **40** 143–53

[13] Li X, Hu X, Yu L, Zhu L, Fu C-W and Heng P-A 2019 CANet: cross-disease attention network for joint diabetic retinopathy and diabetic macular edema grading *IEEE Trans. Med. Imaging* **39** 1483–93

[14] Dekhil O, Naglah A, Shaban M, Ghazal M, Taher F and Elbaz A 2019 Deep learning based method for computer aided diagnosis of diabetic retinopathy *IEEE Int. Conf. on Imaging Systems and Techniques* p 1–4

[15] Luo X, Pu Z, Xu Y, Wong W K, Su J, Dou X, Ye B, Hu J and Mou L 2021 MVDRNet: multi-view diabetic retinopathy detection by combining DCNNs and attention mechanisms *Pattern Recognit.* **120** 108104

[16] Lin Z, Guo R, Wang Y, Wu B, Chen T, Wang W, Chen D Z and Wu J 2018 A framework for identifying diabetic retinopathy based on anti-noise detection and attention-based fusion *Int. Conf. on Medical Image Computing and Computer-Assisted Intervention* pp 74–82

[17] Bodapati J D, Shaik N S and Naralasetti V 2021 Composite deep neural network with gated-attention mechanism for diabetic retinopathy severity classification *J. Ambient Intell. Humaniz. Comput.* **12** 9825–39

[18] Zou J, Zhang X and Lin X 2021 A diabetic retinopathy classification method based on novel attention mechanism *Int. Conf. on Intelligent Computing* pp 129–42

[19] Gondal W M, Köhler J M, Grzeszick R, Fink G A and Hirsch M 2017 Weakly-supervised localization of diabetic retinopathy lesions in retinal fundus images *Int. IEEE Conf. on Image Processing (ICIP)* pp 2069–73

[20] Wang Z and Yang J 2018 Diabetic retinopathy detection via deep convolutional networks for discriminative localization and visual explanation *Workshops at the thirty-second AAAI conference on artificial intelligence*

[21] Kim B *et al* 2018 Interpretability beyond feature attribution: quantitative testing with concept activation vectors (TCAV) *Int. Conf. on Machine Learning* pp 2668–77

[22] Lin C, Zhu J, Shen C, Hu P and Wang Q 2020 ELLG: explainable lesion learning and generation for diabetic retinopathy detection *Int. Joint Conf. on Artificial Intelligence Workshop on Disease Computational Modeling, Int. Joint Conf. on Artificial Intelligence Organization* p 1

[23] Balaratnasingam C, Inoue M, Ahn S, McCann J, Dhrami-Gavazi E, Yannuzzi L A and Freund K B 2016 Visual acuity is correlated with the area of the foveal avascular zone in diabetic retinopathy and retinal vein occlusion *Ophthalmology* **123** 2352–67

[24] Park S S 2016 Cell therapy applications for retinal vascular diseases: diabetic retinopathy and retinal vein occlusion *Investig. Ophthalmol. Vis. Sci.* **57** ORSFj1–10

[25] Pesin N, Mandelcorn E D, Felfeli T, Ogilvie R I and Brent M H 2017 The role of occult hypertension in retinal vein occlusions and diabetic retinopathy *Can. J. Ophthalmol.* **52** S30–3

[26] Tatsumi T, Oshitari T, Baba T, Takatsuna Y and Yamamoto S 2020 Effects of switching from anti-VEGF treatment to triamcinolone acetonide in eyes with refractory macular edema associated with diabetic retinopathy or retinal vein occlusion *BioMed. Res. Int.* **2020** 4529850

[27] Hatef E *et al* 2012 Comparison of time domain and spectral domain optical coherence tomography in measurement of macular thickness in macular edema secondary to diabetic retinopathy and retinal vein occlusion *J. Ophthalmol.* **2012** 354783

[28] Nitta F, Kunikata H, Aizawa N, Omodaka K, Shiga Y, Yasuda M and Nakazawa T 2014 The effect of intravitreal bevacizumab on ocular blood flow in diabetic retinopathy and branch retinal vein occlusion as measured by laser speckle flowgraphy *Clin. Ophthalmol.* **8** 1119

[29] Shin Y U, Kim S, Lee B R, Shin J W and Kim S I 2012 Novel noninvasive detection of the fovea avascular zone using confocal red-free imaging in diabetic retinopathy and retinal vein occlusion *Investig. Ophthalmol. Vis. Sci.* **53** 309–15

[30] Grassmann F, Mengelkamp J, Brandl C, Harsch S, Zimmermann M E, Linkohr B, Peters A, Heid I M, Palm C and Weber B H F 2018 A deep learning algorithm for prediction of age-related eye disease study severity scale for age-related macular degeneration from color fundus photography *Ophthalmology* **125** 1410–20

[31] Shafi A S M *et al* 2021 Statistical texture features based automatic detection and classification of diabetic retinopathy *Proc. Int. Joint Conf. on Advances in Computational Intelligence* pp 27–39

[32] Porwal P, Pachade S, Kamble R, Kokare M, Deshmukh G, Sahasrabuddhe V and Meriaudeau F 2018 Indian diabetic retinopathy image dataset (IDRiD): a database for diabetic retinopathy screening research *Data* **3** 25

[33] Jadhav M L, Shaikh M Z and Sardar V M 2021 Automated microaneurysms detection in fundus images for early diagnosis of diabetic retinopathy *Data Engineering and Intelligent Computing* (Berlin: Springer) pp 87–95

[34] Gonçalves J, Conceiçao T and Soares F 2019 Inter-observer reliability in computer-aided diagnosis of diabetic retinopathy *Proc. 12th Int. Joint Conf. on Biomedical Engineering Systems and Technologies* pp 481–91

[35] Araújo T, Aresta G, Mendonça L, Penas S, Maia C, Carneiro Â, Mendonça A M and Campilho A 2020 DR| GRADUATE: uncertainty-aware deep learning-based diabetic retinopathy grading in eye fundus images *Med. Image Anal.* **63** 101715

[36] Wang S, Wang X, Hu Y, Shen Y, Yang Z, Gan M and Lei B 2020 Diabetic retinopathy diagnosis using multichannel generative adversarial network with semisupervision *IEEE Trans. Autom. Sci. Eng.* **18** 574–85

[37] Li W, Song Y, Chen K, Ying J, Zheng Z, Qiao S, Yang M, Zhang M and Zhang Y 2021 Predictive model and risk analysis for diabetic retinopathy using machine learning: a retrospective cohort study in China *BMJ Open* **11** e050989

[38] Haugen B R *et al* 2016 2015 American thyroid association management guidelines for adult patients with thyroid nodules and differentiated thyroid cancer: the American Thyroid Association guidelines task force on thyroid nodules and differentiated thyroid cancer *Thyroid* **26** 1–133

[39] Islam S M S, Hasan M M and Abdullah S 2018 Deep learning based early detection and grading of diabetic retinopathy using retinal fundus images, arXiv:1812.10595

[40] Navarro P J, Alonso D and Stathis K 2016 Automatic detection of microaneurysms in diabetic retinopathy fundus images using the L* a* b color space *J. Opt. Soc. Am.* A **33** 74–83

[41] Vo H H and Verma A 2016 New deep neural nets for fine-grained diabetic retinopathy recognition on hybrid color space *2016 IEEE Int. Symp. on Multimedia (ISM)* pp 209–15

[42] Kumar P N S, Kumar R R, Sathar A and Sahasranamam V 2013 Automatic detection of exudates in retinal images using histogram analysis *IEEE Recent Advances in Intelligent Computational Systems (RAICS)* pp 277–81

[43] Vo H H and Verma A 2016 Discriminant color texture descriptors for diabetic retinopathy recognition *12th IEEE Int. Conf. on Intelligent Computer Communication and Processing (ICCP)* pp 309–15

[44] Tsao H-Y, Chan P-Y and Su E C-Y 2018 Predicting diabetic retinopathy and identifying interpretable biomedical features using machine learning algorithms *BMC Bioinf.* **19** 111–21

[45] de La Torre J, Valls A and Puig D 2020 A deep learning interpretable classifier for diabetic retinopathy disease grading *Neurocomputing* **396** 465–76

[46] Costa P, Galdran A, Smailagic A and Campilho A 2018 A weakly-supervised framework for interpretable diabetic retinopathy detection on retinal images *IEEE Access* **6** 18747–58

[47] Butt M M, Latif G, Iskandar D N F A, Alghazo J and Khan A H 2019 Multi-channel convolutions neural network based diabetic retinopathy detection from fundus images *Procedia Comput. Sci.* **163** 283–91

[48] Rublee E, Rabaud V, Konolige K and Bradski G 2011 ORB: an efficient alternative to SIFT or SURF *2011 Int. Conf. on Computer Vision* pp 2564–71

[49] Lowe D G 2004 Distinctive image features from scale-invariant keypoints *Int. J. Comput. Vis.* **60** 91–110

[50] Bay H, Tuytelaars T and Van Gool L 2006 SURF: speeded up robust features *European Conf. on Computer Vision* pp 404–17

[51] Oho E, Baba N, Katoh M, Nagatani T, Osumi M, Amako K and Kanaya K 1984 Application of the Laplacian filter to high-resolution enhancement of SEM images *J. Electron Microsc. Tech.* **1** 331–40

[52] Bora D J 2017 A novel approach for color image edge detection using multidirectional Sobel filter on HSV color space *Int. J. Comput. Sci. Eng.* **5** 154–9

[53] Ismail S M, Said L A, Madian A H and Radwan A G 2021 Fractional-order edge detection masks for diabetic retinopathy diagnosis as a case study *Computers* **10** 30

[54] Sayres R *et al* 2019 Using a deep learning algorithm and integrated gradients explanation to assist grading for diabetic retinopathy *Ophthalmology* **126** 552–64

[55] Naglah A, Khalifa F, Khaled R, Abdel Razek A A K, Ghazal M, Giridharan G and El-Baz A 2021 Novel MRI-based CAD system for early detection of thyroid cancer using multi-input CNN *Sensors* **21** 3878

[56] Szegedy C, Ioffe S, Vanhoucke V and Alemi A A 2017 Inception-v4, Inception-ResNet and the impact of residual connections on learning

[57] Szegedy C, Vanhoucke V, Ioffe S, Shlens J and Wojna Z 2016 Rethinking the inception architecture for computer vision *Proc. IEEE Conf. on Computer Vision and Pattern Recognition* pp 2818–26

[58] Vaswani A, Shazeer N, Parmar N, Uszkoreit J, Jones L, Gomez A N, Kaiser and Polosukhin I 2017 Attention is all you need *Adv. Neural Inf. Process. Syst.* **30**

[59] Simonyan K and Zisserman A 2014 Very deep convolutional networks for large-scale image recognition, arXiv:1409.1556

[60] Russakovsky O *et al* 2015 ImageNet large scale visual recognition challenge *Int. J. Comput. Vis.* **115** 211–52

[61] Decencière E *et al* 2014 Feedback on a publicly distributed image database: the Messidor database *Image Anal. Stereol.* **33** 231–4

[62] Ur Rehman M *et al* 2018 Diabetic retinopathy fundus image classification using discrete wavelet transform *2018 2nd Int. Conf. on Engineering Innovation* pp 75–80

[63] Bhaskaranand M, Cuadros J, Ramachandra C, Bhat S, Nittala M G, Sadda S and Solanki K 2015 EyeArt+ EyePACS: automated retinal image analysis for diabetic retinopathy screening in a telemedicine system *Proc. Ophthalmic Medical Image Analysis Int. Workshop* pp 105–12

[64] Davidson I and Ravi S S 2005 Agglomerative hierarchical clustering with constraints: theoretical and empirical results *European Conf. on Principles of Data Mining and Knowledge Discovery* pp 59–70

[65] Elsharkawy M, Sharafeldeen A, Soliman A, Khalifa F, Ghazal M, El-Daydamony E, Atwan A, Sandhu H S and El-Baz A 2022 A novel computer-aided diagnostic system for early detection of diabetic retinopathy using 3D-OCT higher-order spatial appearance model *Diagnostics* **12** 461

[66] Elsharkawy M *et al* 2021 Role of optical coherence tomography imaging in predicting progression of age-related macular disease: a survey *Diagnostics* **11** 2313

[67] Sandhu H S, Elmogy M, Sharafeldeen A T, Elsharkawy M, El-Adawy N, Eltanboly A, Shalaby A, Keynton R and El- Baz A 2020 Automated diagnosis of diabetic retinopathy using clinical biomarkers, optical coherence tomography, and optical coherence tomography angiography *Am. J. Ophthalmol.* **216** 201–6

[68] Sharafeldeen A *et al* 2021 Precise higher-order reflectivity and morphology models for early diagnosis of diabetic retinopathy using OCT images *Sci. Rep.* **11** 4730

[69] Sleman A A, Soliman A, Elsharkawy M, Giridharan G, Ghazal M, Sandhu H, Schaal S, Keynton R, Elmaghraby A and El- Baz A 2021 A novel 3D segmentation approach for extracting retinal layers from optical coherence tomography images *Med. Phys.* **48** 1584–95

[70] Sleman A A, Soliman A, Ghazal M, Sandhu H, Schaal S, Elmaghraby A and El-Baz A 2019 Retinal layers OCT scans 3-D segmentation *IEEE Int. Conf. on Imaging Systems and Techniques* (Piscataway, NJ: IEEE) p 1–6

[71] Eladawi N, Elmogy M, Ghazal M, Helmy O, Aboelfetouh A, Riad A, Schaal S and El-Baz A 2018 Classification of retinal diseases based on OCT images *Front. Biosci.* **23** 247–64

[72] ElTanboly A, Ismail M, Shalaby A, Switala A, El-Baz A, Schaal S, Gimel'farb G and El-Azab M 2017 A computer-aided diagnostic system for detecting diabetic retinopathy in optical coherence tomography images *Med. Phys.* **44** 914–23

[73] Sandhu H S, El-Baz A and Seddon J M 2018 Progress in automated deep learning for macular degeneration *JAMA Ophthalmol.* **136** 1366–7

[74] Ghazal M, Ali S S, Mahmoud A H, Shalaby A M and El-Baz A 2020 Accurate detection of non-proliferative diabetic retinopathy in optical coherence tomography images using convolutional neural networks *IEEE Access* **8** 34387–97

[75] Reda I, Ghazal M, Shalaby A, Elmogy M, AbouEl-Fetouh A, Ayinde B O, AbouEl-Ghar M, Elmaghraby A, Keynton R and El-Baz A 2018 A novel ADCS-based CNN classification system for precise diagnosis of prostate cancer *2018 24th Int. Conf. on Pattern Recognition* (Piscataway, NJ: IEEE) pp 3923–8

[76] Reda I, Khalil A, Elmogy M, Abou El-Fetouh A, Shalaby A, Abou El-Ghar M, Elmaghraby A, Ghazal M and El-Baz A 2018 Deep learning role in early diagnosis of prostate cancer *Technol. Cancer Res. Treat.* **17**

[77] Reda I, Ayinde B O, Elmogy M, Shalaby A, El-Melegy M, El-Ghar M A, El-fetouh A A, Ghazal M and El-Baz A 2018 A new CNN-based system for early diagnosis of prostate cancer *2018 IEEE 15th Int. Symp. on Biomedical Imaging* (Piscataway, NJ) pp 207–10

[78] Ayyad S M *et al* 2022 A new framework for precise identification of prostatic adenocarcinoma *Sensors* **22** 1848

[79] Hammouda K, Khalifa F, El-Melegy M, Ghazal M, Darwish H E, El-Ghar M A and El-Baz A 2021 A deep learning pipeline for grade groups classification using digitized prostate biopsy specimens *Sensors* **21** 6708

[80] Shehata M *et al* 2020 A multimodal computer-aided diagnostic system for precise identification of renal allograft rejection: preliminary results *Med. Phys.* **47** 2427–40

[81] Shehata M, Khalifa F, Soliman A, Ghazal M, Taher F, Abou El-Ghar M, Dwyer A C, Gimel'farb G, Keynton R S and El- Baz A 2018 Computer-aided diagnostic system for early detection of acute renal transplant rejection using diffusion-weighted MRI *IEEE Trans. Biomed. Eng.* **66** 539–52

[82] Hollis E, Shehata M, Abou El-Ghar M, Ghazal M, El-Diasty T, Merchant M, Switala A E and El-Baz A 2017 Statistical analysis of ADCS and clinical biomarkers in detecting acute renal transplant rejection *Br. J. Radiol.* **90** 20170125

[83] Shehata M *et al* 2021 A comprehensive computer-assisted diagnosis system for early assessment of renal cancer tumors *Sensors* **21** 4928

[84] Khalifa F, Beache G M, El-Ghar M A, El-Diasty T, Gimel'farb G, Kong M and El-Baz A 2013 Dynamic contrast-enhanced MRI-based early detection of acute renal transplant rejection *IEEE Trans. Med. Imaging* **32** 1910–27

[85] Khalifa F, El-Ghar M A, Abdollahi B, Frieboes H, El-Diasty T and El-Baz A 2013 A comprehensive non-invasive framework for automated evaluation of acute renal transplant rejection using DCE-MRI *NMR Biomed.* **26** 1460–70

[86] Khalifa F, Elnakib A, Beache G M, Gimel'farb G, El-Ghar M A, Sokhadze G, Manning S, McClure P and El-Baz A 2011 3D kidney segmentation from CT images using a level set approach guided by a novel stochastic speed function *Proc. Int. Conf.*

Medical Image Computing and Computer-Assisted Intervention (Toronto, 18–22 September) pp 587–94

[87] Shehata M, Khalifa F, Hollis E, Soliman A, Hosseini-Asl E, El-Ghar M A, El-Baz M, Dwyer A C, El-Baz A and Keynton R 2016 A new non-invasive approach for early classification of renal rejection types using diffusion-weighted MRI *IEEE Int. Conf. on Image Processing* (Piscataway, NJ: IEEE) pp 136–40

[88] Khalifa F, Soliman A, Takieldeen A, Shehata M, Mostapha M, Shaffie A, Ouseph R, Elmaghraby A and El-Baz A 2016 Kidney segmentation from CT images using a 3D NMF-guided active contour model *IEEE 13th Int. Symp. on Biomedical Imaging* (Piscataway, NJ: IEEE) pp 432–5

[89] Shehata M, Khalifa F, Soliman A, Takieldeen A, El-Ghar M A, Shaffie A, Dwyer A C, Ouseph R, El-Baz A and Keynton R 2016 3D diffusion MRI-based cad system for early diagnosis of acute renal rejection *IEEE 13th Int. Symp. on Biomedical Imaging* (Piscataway, NJ: IEEE) pp 1177–80

[90] Shehata M, Khalifa F, Soliman A, Alrefai R, El-Ghar M A, Dwyer A C, Ouseph R and El-Baz A 2015 A level set-based framework for 3D kidney segmentation from diffusion MR images *IEEE Int. Conf. on Image Processing* (Piscataway, NJ: IEEE) pp 4441–5

[91] Shehata M, Khalifa F, Soliman A, El-Ghar M A, Dwyer A C, Gimel'farb G, Keynton R and El-Baz A 2016 A promising non-invasive CAD system for kidney function assessment *Int. Conf. on Medical Image Computing and Computer-Assisted Intervention* (Berlin: Springer) pp 613–21

[92] Khalifa F, Soliman A, Elmaghraby A, Gimel'farb G and El-Baz A 2017 3D kidney segmentation from abdominal images using spatial-appearance models *Comput. Math. Methods Med.* **2017** 9818506

[93] Hollis E, Shehata M, Khalifa F, El-Ghar M A, El-Diasty T and El-Baz A 2016 Towards non-invasive diagnostic techniques for early detection of acute renal transplant rejection: a review *Egypt. J. Radiol. Nucl. Med.* **48** 257–69

[94] Shehata M, Khalifa F, Soliman A, El-Ghar M A, Dwyer A C and El-Baz A 2017 Assessment of renal transplant using image and clinical-based biomarkers *Proc. 13th Annual Scientific Meeting of American Society for Diagnostics and Interventional Nephrology (New Orleans, LA, 10–12 February)*

[95] Shehata M, Khalifa F, Soliman A, El-Ghar M A, Dwyer A C and El-Baz A 2016 Early assessment of acute renal rejection *Proc. 12th Annual Scientific Meeting of American Society for Diagnostics and Interventional Nephrology (Phoenix, AZ, 19–21 February)*

[96] Eltanboly A, Ghazal M, Hajjdiab H, Shalaby A, Switala A, Mahmoud A, Sahoo P, El-Azab M and El-Baz A 2019 Level sets-based image segmentation approach using statistical shape priors *Appl. Math. Comput.* **340** 164–79

[97] Shehata M, Mahmoud A, Soliman A, Khalifa F, Ghazal M, El-Ghar M A, El-Melegy M and El-Baz A 2018 3D kidney segmentation from abdominal diffusion MRI using an appearance-guided deformable boundary *PLoS One* **13** e0200082

[98] Abdeltawab H *et al* 2019 A novel CNN-based cad system for early assessment of transplanted kidney dysfunction *Sci. Rep.* **9** 5948

[99] Hammouda K *et al* 2020 A new framework for performing cardiac strain analysis from cine MRI imaging in mice *Sci. Rep.* **10** 7725

[100] Abdeltawab H, Khalifa F, Hammouda K, Miller J M, Meki M M, Ou Q, El-Baz A and Mohamed T 2021 Artificial intelligence based framework to quantify the cardiomyocyte structural integrity in heart slices *Cardiovasc. Eng. Technol.* **13** 170–80

[101] Khalifa F, Beache G M, Elnakib A, Sliman H, Gimel'farb G, Welch K C and El-Baz A 2013 A new shape-based framework for the left ventricle wall segmentation from cardiac first-pass perfusion MRI *Proc. IEEE Int. Symp. on Biomedical Imaging: From Nano to Macro (San Francisco, CA)* pp 41–4

[102] Khalifa F, Beache G M, Elnakib A, Sliman H, Gimel'farb G, Welch K C and El-Baz A 2012 A new nonrigid registration framework for improved visualization of transmural perfusion gradients on cardiac first-pass perfusion MRI *Proc. IEEE Int. Symp. on Biomedical Imaging: From Nano to Macro (Barcelona)* pp 828–31

[103] Khalifa F, Beache G M, Firjani A, Welch K C, Gimel'farb G and El-Baz A 2012 A new nonrigid registration approach for motion correction of cardiac first-pass perfusion MRI *Proc. IEEE Int. Conf. on Image Processing (Lake Buena Vista, FL, 30 September–3 October)* pp 1665–8

[104] Khalifa F, Beache G M, Gimel'farb G and El-Baz A 2012 A novel CAD system for analyzing cardiac first-pass MR images *Proc. IAPR Int. Conf. on Pattern Recognition (Tsukuba Science City)* pp 77–80

[105] Khalifa F, Beache G M, Gimel'farb G and El-Baz A 2011 A novel approach for accurate estimation of left ventricle global indexes from short-axis cine MRI *Proc. IEEE Int. Conf. on Image Processing (Brussels)* pp 2645–9

[106] Khalifa F, Beache G M, Gimel'farb G, Giridharan G A and El-Baz A 2011 A new image-based framework for analyzing cine images *Handbook of Multi Modality State-of-the-Art Medical Image Segmentation and Registration Methodologies* ed A El-Baz, U R Acharya, M Mirmedhdi and J S Suri vol 2 (New York: Springer) pp 69–98

[107] Khalifa F, Beache G M, Gimel'farb G, Giridharan G A and El-Baz A 2012 Accurate automatic analysis of cardiac cine images *IEEE Trans. Biomed. Eng.* **59** 445–55

[108] Khalifa F, Beache G M, Nitzken M, Gimel'farb G, Giridharan G A and El-Baz A 2011 Automatic analysis of left ventricle wall thickness using short-axis cine CMR images *Proc. IEEE Int. Symp. on Biomedical Imaging: From Nano to Macro (Chicago, IL, 30 March–2 April)* pp 1306–9

[109] Nitzken M, Beache G, Elnakib A, Khalifa F, Gimel'farb G and El-Baz A 2012 Accurate modeling of tagged CMR 3D image appearance characteristics to improve cardiac cycle strain estimation *19th IEEE Int. Conf. on Image Processing (Orlando, FL, September)* (Piscataway, NJ: IEEE) pp 521–4

[110] Nitzken M, Beache G, Elnakib A, Khalifa F, Gimel'farb G and El-Baz A 2012 Improving full-cardiac cycle strain estimation from tagged CMR by accurate modeling of 3D image appearance characteristics *9th IEEE Int. Symp. on Biomedical Imaging (Barcelona, May)* (Piscataway, NJ: IEEE) pp 462–5

[111] Nitzken M J, El-Baz A S and Beache G M 2012 Markov–Gibbs random field model for improved full-cardiac cycle strain estimation from tagged CMR *J. Cardiovasc. Magn. Res.* **14** P258

[112] Sliman H, Elnakib A, Beache G, Elmaghraby A and El-Baz A 2014 Assessment of myocardial function from cine cardiac MRI using a novel 4D tracking approach *J. Comput. Sci. Syst. Biol.* **7** 169–73

[113] Sliman H, Elnakib A, Beache G M, Soliman A, Khalifa F, Gimel'farb G, Elmaghraby A and El-Baz A 2014 A novel 4D PDE-based approach for accurate assessment of myocardium function using cine cardiac magnetic resonance images *Proc. IEEE Int. Conf. on Image Processing (Paris, 27–30 October)* pp 3537–41

[114] Sliman H, Khalifa F, Elnakib A, Beache G M, Elmaghraby A and El-Baz A 2013 A new segmentation-based tracking framework for extracting the left ventricle cavity from cine cardiac MRI *Proc. IEEE Int. Conf. on Image Processing (Melbourne, 15–18 September)* pp 685–9

[115] Sliman H, Khalifa F, Elnakib A, Soliman A, Beache G M, Elmaghraby A, Gimel'farb G and El-Baz A 2013 Myocardial borders segmentation from cine MR images using bi-directional coupled parametric deformable models *Med. Phys.* **40** 092302

[116] Sliman H, Khalifa F, Elnakib A, Soliman A, Beache G M, Gimel'farb G, Emam A, Elmaghraby A and El-Baz A 2013 Accurate segmentation framework for the left ventricle wall from cardiac cine MRI *Proc. Int. Symp. on Computational Models for Life Science* vol 1559 *(Sydney, 27–29 November)* pp 287–96

[117] Sharafeldeen A, Elsharkawy M, Alghamdi N S, Soliman A and El-Baz A 2021 Precise segmentation of Covid-19 infected lung from CT images based on adaptive first-order appearance model with morphological/anatomical constraints *Sensors* **21** 5482

[118] Elsharkawy M *et al* 2021 Early assessment of lung function in coronavirus patients using invariant markers from chest x-rays images *Sci. Rep.* **11** 12095

[119] Abdollahi B, Civelek A C, Li X-F, Suri J and El-Baz A 2014 PET/CT nodule segmentation and diagnosis: a survey *Multi Detector CT Imaging* ed L Saba and J S Suri (Boca Raton, FL: Taylor and Francis) ch 30 pp 639–51

[120] Abdollahi B, El-Baz A and Amini A A 2011 A multi-scale non-linear vessel enhancement technique *2011 Annual Int. Conf. of the IEEE Engineering in Medicine and Biology Society* (Piscataway, NJ: IEEE) pp 3925–9

[121] Abdollahi B, Soliman A, Civelek A, Li X-F, Gimel'farb G and El-Baz A 2012 A novel Gaussian scale space-based joint MGRF framework for precise lung segmentation *Proc. IEEE Int. Conf. on Image Processing* (Piscataway, NJ: IEEE) pp 2029–32

[122] Abdollahi B, Soliman A, Civelek A, Li X-F, Gimel'farb G and El-Baz A 2012 A novel 3D joint MGRF framework for precise lung segmentation *Machine Learning in Medical Imaging* (Berlin: Springer) pp 86–93

[123] Ali A M, El-Baz A S and Farag A A 2007 A novel framework for accurate lung segmentation using graph cuts *Proc. IEEE Int. Symp. on Biomedical Imaging: From Nano to Macro* (Piscataway, NJ: IEEE) pp 908–11

[124] El-Baz A, Beache G M, Gimel'farb G, Suzuki K and Okada K 2013 Lung imaging data analysis *Int. J. Biomed. Imaging* **2013** 618561

[125] El-Baz A, Beache G M, Gimel'farb G, Suzuki K, Okada K, Elnakib A, Soliman A and Abdollahi B 2013 Computer-aided diagnosis systems for lung cancer: challenges and methodologies *Int. J. Biomed. Imaging* **2013** 942353

[126] El-Baz A, Elnakib A, Abou El-Ghar M, Gimel'farb G, Falk R and Farag A 2013 Automatic detection of 2D and 3D lung nodules in chest spiral CT scans *Int. J. Biomed. Imaging* **2013** 517632

[127] El-Baz A, Farag A A, Falk R and La Rocca R 2003 A unified approach for detection, visualization, and identification of lung abnormalities in chest spiral CT scans *Int. Congress Series* **1256** 998–1004

[128] El-Baz A, Farag A A, Falk R and La Rocca R 2002 Detection, visualization and identification of lung abnormalities in chest spiral CT scan: phase-I *Proc. Int. Conf. on Biomedical Engineering* vol 12 *(Cairo, Egypt)*

[129] El-Baz A, Farag A, Gimel'farb G, Falk R, El-Ghar M A and Eldiasty T 2006 A framework for automatic segmentation of lung nodules from low dose chest CT scans *Proc. Int. Conf. on Pattern Recognition* vol 3 (Piscataway, NJ: IEEE) pp 611–4

[130] El-Baz A, Farag A, Gimel'farb G, Falk R and El-Ghar M A 2011 A novel level set-based computer-aided detection system for automatic detection of lung nodules in low dose chest computed tomography scans *Lung Imag. Comput. Aided Diagn.* **10** 221–38

[131] El-Baz A, Gimel'farb G, Abou M, El-Ghar R and Falk 2012 Appearance-based diagnostic system for early assessment of malignant lung nodules *Proc. IEEE Int. Conf. on Image Processing* (Piscataway, NJ: IEEE) pp 533–6

[132] El-Baz A, Gimel'farb G and Falk R 2011 A novel 3D framework for automatic lung segmentation from low dose CT images *Lung Imaging and Computer Aided Diagnosis* ed A El-Baz and J S Suri (Boca Raton, FL: Taylor and Francis) pp 1–16 ch 1

[133] El-Baz A, Gimel'farb G, Falk R and El-Ghar M 2010 Appearance analysis for diagnosing malignant lung nodules *Proc. IEEE Int. Symp. on Biomedical Imaging: From Nano to Macro* (Piscataway, NJ: IEEE) pp 193–6

[134] El-Baz A, Gimel'farb G, Falk R and El-Ghar M A 2011 A novel level set-based CAD system for automatic detection of lung nodules in low dose chest CT scans *Lung Imaging and Computer Aided Diagnosis* ed El-Baz and J S Suri vol 1 (Boca Raton, FL: Taylor and Francis) pp 221–38 ch 10

[135] El-Baz A, Gimel'farb G, Falk R and El-Ghar M A 2008 A new approach for automatic analysis of 3D low dose CT images for accurate monitoring the detected lung nodules *Proc. Int. Conf. on Pattern Recognition* (Piscataway, NJ: IEEE) pp 1–4

[136] El-Baz A, Gimel'farb G, Falk R and El-Ghar M A 2007 A novel approach for automatic follow-up of detected lung nodules *Proc. IEEE Int. Conf. on Image Processing* vol 5 (Piscataway, NJ: IEEE)

[137] El-Baz A, Gimel'farb G, Falk R and El-Ghar M A 2007 A new CAD system for early diagnosis of detected lung nodules *IEEE Int. Conf. on Image Processing* vol 2 (Piscataway, NJ: IEEE)

[138] El-Baz A, Gimel'farb G, Falk R, El-Ghar M A and Refaie H 2008 Promising results for early diagnosis of lung cancer *Proc. IEEE Int. Symp. on Biomedical Imaging: From Nano to Macro* (Piscataway, NJ: IEEE) pp 1151–4

[139] El-Baz A, Gimel'farb G L, Falk R, Abou El-Ghar M, Holland T and Shaffer T 2008 A new stochastic framework for accurate lung segmentation *Proc. Medical Image Computing and Computer-Assisted Intervention* pp 322–30

[140] El-Baz A, Gimel'farb G L, Falk R, Heredis D and Abou El-Ghar M 2008 A novel approach for accurate estimation of the growth rate of the detected lung nodules *Proc. Int. Workshop on Pulmonary Image Analysis* pp 33–42

[141] El-Baz A, Gimel'farb G L, Falk R, Holland T and Shaffer T 2008 A framework for unsupervised segmentation of lung tissues from low dose computed tomography images *Proc. British Machine Vision* pp 1–10

[142] El-Baz A, Gimel'farb G, Falk R and El-Ghar M A 2011 3D MGRF-based appearance modeling for robust segmentation of pulmonary nodules in 3D LDCT chest images *Lung Imaging and Computer Aided Diagnosis* (Boca Raton, FL: Taylor and Francis) ch 3 pp 51–63

[143] El-Baz A, Gimel'farb G, Falk R and El-Ghar M A 2009 Automatic analysis of 3D low dose CT images for early diagnosis of lung cancer *Pattern Recognit.* **42** 1041–51

[144] El-Baz A, Gimel'farb G, Falk R, El-Ghar M A, Rainey S, Heredia D and Shaffer T 2009 Toward early diagnosis of lung cancer *Proc. Medical Image Computing and Computer-Assisted Intervention* (Berlin: Springer) pp 682–9

[145] El-Baz A, Gimel'farb G, Falk R, El-Ghar M A and Suri J 2011 Appearance analysis for the early assessment of detected lung nodules *Lung Imaging and Computer Aided Diagnosis* (Boca Raton, FL: Taylor and Francis) ch 17 pp 395–404

[146] El-Baz A, Khalifa F, Elnakib A, Nitkzen M, Soliman A, McClure P, Gimel'farb G and El-Ghar M A 2012 A novel approach for global lung registration using 3D Markov Gibbs appearance model *Proc. Int. Conf. Medical Image Computing and Computer-Assisted Intervention (Nice, 1–5 October)* pp 114–21

[147] El-Baz A, Nitzken M, Elnakib A, Khalifa F, Gimel'farb G, Falk R and El-Ghar M A 2011 3D shape analysis for early diagnosis of malignant lung nodules *Proc. Int. Conf. Medical Image Computing and Computer-Assisted Intervention (Toronto, 18–22 September)* pp 175–82

[148] El-Baz A, Nitzken M, Gimel'farb G, Van Bogaert E, Falk R, El-Ghar M A and Suri J 2011 Three-dimensional shape analysis using spherical harmonics for early assessment of detected lung nodules *Lung Imaging and Computer Aided Diagnosis* (Boca Raton, FL: Taylor and Francis) ch 19 pp 421–38

[149] El-Baz A, Nitzken M, Khalifa F, Elnakib A, Gimel'farb G, Falk R and El-Ghar M A 2011 3D shape analysis for early diagnosis of malignant lung nodules *Proc. Int. Conf. on Information Processing in Medical Imaging (Monastery Irsee, 3–8 July)* pp 772–83

[150] El-Baz A, Nitzken M, Vanbogaert E, Gimel'Farb G, Falk R and El-Ghar M Abo 2011 A novel shape-based diagnostic approach for early diagnosis of lung nodules *IEEE Int. Symp. on Biomedical Imaging: From Nano to Macro* (Piscataway, NJ: IEEE) pp 137–40

[151] El-Baz A, Sethu P, Gimel'farb G, Khalifa F, Elnakib A, Falk R and El-Ghar M A 2011 Elastic phantoms generated by microfluidics technology: validation of an imaged-based approach for accurate measurement of the growth rate of lung nodules *Biotechnol. J.* **6** 195–203

[152] El-Baz A, Sethu P, Gimel'farb G, Khalifa F, Elnakib A, Falk R and El-Ghar M A 2010 A new validation approach for the growth rate measurement using elastic phantoms generated by state-of-the-art microfluidics technology *Proc. IEEE Int. Conf. on Image Processing (Hong Kong, 26–29 September)* pp 4381–3

[153] El-Baz A, Sethu P, Gimel'farb G, Khalifa F, Elnakib A, Falk R and Suri M A E-G J 2011 Validation of a new imaged-based approach for the accurate estimating of the growth rate of detected lung nodules using real CT images and elastic phantoms generated by state-of-the-art microfluidics technology *Handbook of Lung Imaging and Computer Aided Diagnosis* ed A El-Baz and J S Suri vol 1 (New York: Taylor and Francis) ch 18 pp 405–20

[154] El-Baz A, Soliman A, McClure P, Gimel'farb G, El-Ghar M A and Falk R 2012 Early assessment of malignant lung nodules based on the spatial analysis of detected lung nodules *Proc. IEEE Int. Symp. on Biomedical Imaging: From Nano to Macro* (Piscataway, NJ: IEEE) pp 1463–6

[155] El-Baz A, Yuksel S E, Elshazly S and Farag A A 2005 Non-rigid registration techniques for automatic follow-up of lung nodules *Proc. Computer Assisted Radiology and Surgery* vol 1281 (Amsterdam: Elsevier) pp 1115–20

[156] El-Baz A S and Suri J S 2011 *Lung Imaging and Computer Aided Diagnosis* (Boca Raton, FL: CRC Press)

[157] Soliman A, Khalifa F, Dunlap N, Wang B, El-Ghar M and El-Baz A 2016 An ISO-surfaces based local deformation handling framework of lung tissues *13th IEEE Int. Symp. on Biomedical Imaging* (Piscataway, NJ: IEEE) pp 1253–9

[158] Soliman A, Khalifa F, Shaffie A, Dunlap N, Wang B, Elmaghraby A and El-Baz A 2016 Detection of lung injury using 4D-CT chest images *13th IEEE Int. Symp. on Biomedical Imaging* (Piscataway, NJ: IEEE) pp 1274–7

[159] Soliman A, Khalifa F, Shaffie A, Dunlap N, Wang B, Elmaghraby A, Gimel'farb G, Ghazal M and El-Baz A 2017 A comprehensive framework for early assessment of lung injury *IEEE Int. Conf. on Image Processing* (Piscataway, NJ: IEEE) pp 3275–9

[160] Shaffie A, Soliman A, Ghazal M, Taher F, Dunlap N, Wang B, Elmaghraby A, Gimel'farb G and El-Baz A 2017 A new framework for incorporating appearance and shape features of lung nodules for precise diagnosis of lung cancer *IEEE Int. Conf. on Image Processing* (Piscataway, NJ: IEEE) pp 1372–6

[161] Soliman A, Khalifa F, Shaffie A, Liu N, Dunlap N, Wang B, Elmaghraby A, Gimel'farb G and El-Baz A 2016 Image-based CAD system for accurate identification of lung injury *IEEE Int. Conf. on Image Processing* (Piscataway, NJ: IEEE) pp 121–5

[162] Soliman A, Shaffie A, Ghazal M, Gimel'farb G, Keynton R and El-Baz A 2018 A novel CNN segmentation framework based on using new shape and appearance features *25th IEEE Int. Conf. on Image Processing* (Piscataway, NJ: IEEE) pp 3488–92

[163] Shaffie A, Soliman A, Khalifeh H A, Ghazal M, Taher F, Keynton R, Elmaghraby A and El-Baz A 2018 On the integration of CT-derived features for accurate detection of lung cancer *IEEE Int. Symp. on Signal Processing and Information Technology* (Piscataway, NJ: IEEE) pp 435–40

[164] Shaffie A, Soliman A, Khalifeh H A, Ghazal M, Taher F, Elmaghraby A, Keynton R and El-Baz A 2019 Radiomic-based framework for early diagnosis of lung cancer *16th IEEE Int. Symp. on Biomedical Imaging* (Piscataway, NJ: IEEE) pp 1293–7

[165] Shaffie A, Soliman A, Ghazal M, Taher F, Dunlap N, Wang B, Van Berkel V, Gimelfarb G, Elmaghraby A and El-Baz A 2018 A novel autoencoder-based diagnostic system for early assessment of lung cancer *25th IEEE Int. Conf. on Image Processing* (Piscataway, NJ: IEEE) pp 1393–7

[166] Shaffie A *et al* 2018 A generalized deep learning-based diagnostic system for early diagnosis of various types of pulmonary nodules *Technol. Cancer Res. Treat.* **17** 1533033818798800

[167] Abdel Razek A A K, Alksas A, Shehata M, AbdelKhalek A, Abdel Baky K, El-Baz A and Helmy E 2021 Clinical applications of artificial intelligence and radiomics in neuro-oncology imaging *Insights Imaging* **12** 152

[168] ElNakieb Y *et al* 2018 Towards accurate personalized autism diagnosis using different imaging modalities: SMRI, FMRI, and DTI *IEEE Int. Symp. on Signal Processing and Information Technology* (Piscataway, NJ: IEEE) pp 447–52

[169] ElNakieb Y *et al* 2019 Autism spectrum disorder diagnosis framework using diffusion tensor imaging *IEEE Int. Conf. on Imaging Systems and Techniques* (Piscataway, NJ: IEEE) pp 1–5

[170] Haweel R, Dekhil O, Shalaby A, Mahmoud A, Ghazal M, Keynton R, Barnes G and El-Baz A 2019 A machine learning approach for grading autism severity levels using

task-based functional MRI *IEEE Int. Conf. on Imaging Systems and Techniques* (Piscataway, NJ: IEEE) pp 1–5

[171] Dekhil O *et al* 2020 A comprehensive framework for differentiating autism spectrum disorder from neurotypicals by fusing structural MRI and resting state functional MRI *Seminars Pediat. Neurol.* **34** 100805

[172] Haweel R, Dekhil O, Shalaby A, Mahmoud A, Ghazal M, Khalil A, Keynton R, Barnes G and El-Baz A 2020 A novel framework for grading autism severity using task-based FMRI *17th IEEE Int. Symp. on Biomedical Imaging* (Piscataway, NJ: IEEE) pp 1404–7

[173] El-Baz A, Elnakib A, Khalifa F, El-Ghar M A, McClure P, Soliman A and Gimel'farb G 2012 Precise segmentation of 3-D magnetic resonance angiography *IEEE Trans. Biomed. Eng.* **59** 2019–29

[174] El-Baz A, Farag A, Elnakib A, Casanova M F, Gimel'farb G, Switala A E, Jordan D and Rainey S 2011 Accurate automated detection of autism related corpus callosum abnormalities *J. Med. Syst.* **35** 929–39

[175] El-Baz A, Gimel'farb G, Falk R, El-Ghar M A, Kumar V and Heredia D 2009 A novel 3D joint Markov–Gibbs model for extracting blood vessels from PC–MRA images *Proc. Conf. Medical Image Computing and Computer-Assisted Intervention* vol 5762 (Berlin: Springer) pp 943–50

[176] Elnakib A, El-Baz A, Casanova M F, Gimel'farb G and Switala A E 2010 Image-based detection of corpus callosum variability for more accurate discrimination between dyslexic and normal brains *Proc. IEEE Int. Symp. on Biomedical Imaging: From Nano to Macro* (Piscataway, NJ: IEEE) pp 109–12

[177] Elnakib A, Casanova M F, Gimel'farb G, Switala A E and El-Baz A 2011 Autism diagnostics by centerline-based shape analysis of the corpus callosum *Proc. IEEE Int. Symp. on Biomedical Imaging: From Nano to Macro* (Piscataway, NJ: IEEE) pp 1843–6

[178] Elnakib A, Nitzken M, Casanova M, Park H, Gimel'farb G and El-Baz A 2012 Quantification of age-related brain cortex change using 3D shape analysis *21st Int. Conf. on Pattern Recognition* (Piscataway, NJ: IEEE) pp 41–4

[179] Nitzken M, Casanova M, Gimel'farb G, Elnakib A, Khalifa F, Switala A and El-Baz A 2011 3D shape analysis of the brain cortex with application to dyslexia *18th IEEE Int. Conf. on Image Processing (Brussels, September)* (Piscataway, NJ: IEEE) pp 2657–60

[180] El-Gamal F E-Z A, Elmogy M M, Ghazal M, Atwan A, Barnes G N, Casanova M F, Keynton R and El-Baz A S 2017 A novel CAD system for local and global early diagnosis of Alzheimer's disease based on PIB-PET scans *IEEE Int. Conf. on Image Processing* (Piscataway, NJ: IEEE) pp 3270–4

[181] Ismail M M, Keynton R S, Mostapha M M, ElTanboly A H, Casanova M F, Gimel'farb G L and El-Baz A 2016 Studying autism spectrum disorder with structural and diffusion magnetic resonance imaging: a survey *Front. Human Neurosci.* **10** 211

[182] Alansary A *et al* 2016 Infant brain extraction in T1-weighted MR images using BET and refinement using LCDG and MGRF models *IEEE J. Biomed. Health Inform.* **20** 925–35

[183] Asl E H, Ghazal M, Mahmoud A, Aslantas A, Shalaby A, Casanova M, Barnes G, Gimel'farb G, Keynton R and El-Baz A 2018 Alzheimer's disease diagnostics by a 3D deeply supervised adaptable convolutional network *Front. Biosci.* **23** 584–96

[184] Dekhil O *et al* 2019 A personalized autism diagnosis cad system using a fusion of structural MRI and resting-state functional MRI data *Front. Psychiatry* **10** 392

[185] Dekhil O, Shalaby A, Soliman A, Mahmoud A, Kong M, Barnes G, Elmaghraby A and El-Baz A 2021 Identifying brain areas correlated with ADOS raw scores by studying altered dynamic functional connectivity patterns *Med. Image Anal.* **68** 101899

[186] Elnakieb Y A *et al* 2020 Computer aided autism diagnosis using diffusion tensor imaging *IEEE Access* **8** 191298–308

[187] Ali M T, Elnakieb Y A, Shalaby A, Mahmoud A, Switala A, Ghazal M, Khelifi A, Fraiwan L, Barnes G and El-Baz A 2021 Autism classification using SMRI: a recursive features selection based on sampling from multi-level high dimensional spaces *18th IEEE Int. Symp. on Biomedical Imaging* (Piscataway, NJ: IEEE) pp 267–70

[188] Ali M T *et al* 2022 The role of structure MRI in diagnosing autism *Diagnostics* **12** 165

[189] ElNakieb Y, Ali M T, Elnakib A, Shalaby A, Soliman A, Mahmoud A, Ghazal M, Barnes G N and El-Baz A 2021 The role of diffusion tensor MR imaging (DTI) of the brain in diagnosing autism spectrum disorder: promising results *Sensors* **21** 8171

[190] Mahmoud A, El-Barkouky A, Farag H, Graham J and Farag A 2013 A non-invasive method for measuring blood flow rate in superficial veins from a single thermal image *Proc. IEEE Conf. on Computer Vision and Pattern Recognition Workshops* pp 354–9

[191] Elsaid N *et al* 2021 Impact of stress and hypertension on the cerebrovasculature *Front. Biosci.* **26** 1643

[192] Taher F, Kandil H, Gebru Y, Mahmoud A, Shalaby A, El-Mashad S and El-Baz A 2021 A novel MRA-based framework for segmenting the cerebrovascular system and correlating cerebral vascular changes to mean arterial pressure *Appl. Sci.* **11** 4022

[193] Kandil H, Soliman A, Taher F, Ghazal M, Khalil A, Giridharan G, Keynton R, Jennings J R and El-Baz A 2020 A novel computer-aided diagnosis system for the early detection of hypertension based on cerebrovascular alterations *NeuroImage: Clin.* **25** 102107

[194] Kandil H, Soliman A, Ghazal M, Mahmoud A, Shalaby A, Keynton R, Elmaghraby A, Giridharan G and El-Baz A 2019 A novel framework for early detection of hypertension using magnetic resonance angiography *Sci. Rep.* **9** 11105

[195] Gebru Y, Giridharan G, Ghazal M, Mahmoud A, Shalaby A and El-Baz A 2018 Detection of cerebrovascular changes using magnetic resonance angiography *Cardiovascular Imaging and Image Analysis* (Boca Raton, FL: CRC Press) pp 1–22

[196] Mahmoud A, Shalaby A, Taher F, El-Baz M, Suri J S and El-Baz A 2018 Vascular tree segmentation from different image modalities *Cardiovascular Imaging and Image Analysis* (Boca Raton, FL: CRC Press) pp 43–70

[197] Taher F, Mahmoud A, Shalaby A and El-Baz A 2018 A review on the cerebrovascular segmentation methods *IEEE Int. Symp. on Signal Processing and Information Technology* (Piscataway, NJ: IEEE) pp 359–64

[198] Kandil H, Soliman A, Fraiwan L, Shalaby A, Mahmoud A, ElTanboly A, Elmaghraby A, Giridharan G and El-Baz A 2018 A novel MRA framework based on integrated global and local analysis for accurate segmentation of the cerebral vascular system *15th IEEE Int. Symp. on Biomedical Imaging* (Piscataway, NJ: IEEE) pp 1365–8

[199] Taher F, Soliman A, Kandil H, Mahmoud A, Shalaby A, Gimel'farb G and El-Baz A 2020 Accurate segmentation of cerebrovasculature from TOF-MRA images using appearance descriptors *IEEE Access* **8** 96139–49

[200] Taher F, Soliman A, Kandil H, Mahmoud A, Shalaby A, Gimel'farb G and El-Baz A 2020 Precise cerebrovascular segmentation *IEEE Int. Conf. on Image Processing* (Piscataway, NJ: IEEE) pp 394–7

[201] Hammouda K *et al* 2019 A CNN-based framework for bladder wall segmentation using MRI *Fifth Int. Conf. on Advances in Biomedical Engineering* (Piscataway, NJ: IEEE) pp 1–4

[202] Hammouda K, Khalifa F, Soliman A, Ghazal M, Abou El-Ghar M, Haddad A, Elmogy M, Darwish H, Keynton R and El-Baz A 2019 A deep learning-based approach for accurate segmentation of bladder wall using MR images *IEEE Int. Conf. on Imaging Systems and Techniques* (Piscataway, NJ: IEEE) pp 1–6

[203] Hammouda K, Khalifa F, Soliman A, Abdeltawab H, Ghazal M, Abou El-Ghar M, Haddad A, Darwish H E, Keynton R and El-Baz A 2020 A 3D CNN with a learnable adaptive shape prior for accurate segmentation of bladder wall using MR images *17th IEEE Int. Symp. on Biomedical Imaging* (Piscataway, NJ: IEEE) pp 935–8

[204] Hammouda K, Khalifa F, Soliman A, Ghazal M, Abou El-Ghar M, Badawy M, Darwish H, Khelifi A and El-Baz A 2021 A multiparametric MRI-based CAD system for accurate diagnosis of bladder cancer staging *Comput. Med. Imaging Graph.* **90** 101911

[205] Hammouda K, Khalifa F, Soliman A, Ghazal M, Abou El-Ghar M, Badawy M, Darwish H and El-Baz A 2021 A CAD system for accurate diagnosis of bladder cancer staging using a multiparametric MRI *18th IEEE Int. Symp. on Biomedical Imaging* (Piscataway, NJ: IEEE) pp 1718–21

[206] Alksas A, Shehata M, Saleh G A, Shaffie A, Soliman A, Ghazal M, Khalifeh H A, Razek A A and El-Baz A 2021 A novel computer-aided diagnostic system for early assessment of hepatocellular carcinoma *25th Int. Conf. on Pattern Recognition* (Piscataway, NJ: IEEE) pp 10375–82

[207] Alksas A *et al* 2021 A novel computer-aided diagnostic system for accurate detection and grading of liver tumors *Sci. Rep.* **11** 1–18

[208] Razek A A K A, Khaled R, Helmy E, Naglah A, AbdelKhalek A and El-Baz A 2022 Artificial intelligence and deep learning of head and neck cancer *Magn. Reson. Imaging Clin.* **30** 81–94

[209] Sharafeldeen A *et al* 2021 Texture and shape analysis of diffusion-weighted imaging for thyroid nodules classification using machine learning *Med. Phys.* **49** 988–9

[210] Naglah A, Khalifa F, Mahmoud A, Ghazal M, Jones P, Murray T, Elmaghraby A S and El-Baz A 2018 Athlete-customized injury prediction using training load statistical records and machine learning *IEEE Int. Symp. on Signal Processing and Information Technology* (Piscataway, NJ: IEEE) pp 459–64

[211] Mahmoud A H 2014 Utilizing radiation for smart robotic applications using visible, thermal, and polarization images *PhD thesis* University of Louisville, KY

[212] Mahmoud A, El-Barkouky A, Graham J and Farag A 2014 Pedestrian detection using mixed partial derivative based histogram of oriented gradients *IEEE Int. Conf. on Image Processing* (Piscataway, NJ: IEEE) pp 2334–7

[213] El-Barkouky A, Mahmoud A, Graham J and Farag A 2013 An interactive educational drawing system using a humanoid robot and light polarization *IEEE Int. Conf. on Image Processing* (Piscataway, NJ: IEEE) pp 3407–11

[214] Mahmoud A H, El-Melegy M T and Farag A A 2012 Direct method for shape recovery from polarization and shading *19th IEEE Int. Conf. on Image Processing* (Piscataway, NJ: IEEE) pp 1769–72

[215] Ghazal M A, Mahmoud A, Aslantas A, Soliman A, Shalaby A, Benediktsson J A and El-Baz A 2019 Vegetation cover estimation using convolutional neural networks *IEEE Access* **7** 132563–76

[216] Ghazal M, Mahmoud A, Shalaby A and El-Baz A 2019 Automated framework for accurate segmentation of leaf images for plant health assessment *Environ. Monit. Assess.* **191** 491

[217] Ghazal M, Mahmoud A, Shalaby A, Shaker S, Khelifi A and El-Baz A 2020 Precise statistical approach for leaf segmentation *IEEE Int. Conf. on Image Processing* (Piscataway, NJ: IEEE) pp 2985–9

IOP Publishing

Photo Acoustic and Optical Coherence Tomography Imaging, Volume 2

Fundus imaging for the retina

Ayman El-Baz and Jasjit S Suri

Chapter 2

A two-phase novel optic disc detection algorithm based on vesselness distribution and a fuzzy classifier using vessel ramification and fundus color features

Mehmet Nergiz

2.1 Introduction

The optic disc (OD) is a basic landmark for retinal anatomy whose location can be used for detecting the other retinal landmarks such as blood vessels and macula [1, 2]. A change in the cup-to-disc ratio of the OD is used for the diagnosis of glaucoma. The shape of the OD can be used to roughly diagnose optic neuritis, optic neuropathies, brain tumor, papilledema, and cardiovascular diseases [3]. Ophthalmologists can limit the consequences of retinal diseases such as macular edema, glaucoma, diabetic retinopathy (DR), and blindness by detecting anomalies and lesions such as exudates, aneurysms, hemorrhages, and change in the shape and size of the OD [4]. The World Health Organization has published figures that show that the number of people with diabetes rose to 422 million in 2014 [5, 6].

Automated retinal analysis software systems have a medical and commercial value by aiming to dramatically decrease the cost of clinical eye exams [7]. These kinds of automated analysis systems can be helpful, particularly for developing countries which suffer from an insufficient number of ophthalmologists [8, 9]. The retinal image systems not only help in the diagnosis of diseases or segmentation of certain anatomic structures, but also in yielding some measurements which can be used as biomarkers and decrease intra- and inter-observer variation [7, 10, 11].

The landmark property of the OD is also beneficial in terms of estimating the distance of the lesions, edema, and hemorrhages from macula and OD [4]. The bright and yellowish color of the OD can act as a confounder in the retinal image

doi:10.1088/978-0-7503-2056-6ch2

analysis or AI systems which aim to detect or differentiate exudates. Thus, locating and excluding the OD can help in these kinds of exudate or DR detection systems [2, 4, 12]. The OD localization is also a mandatory step for the arteriolar-to-venular-ratio (AVR) index whose value is used as a biomarker for stroke and cerebral atrophy [7, 13].

The literature of OD detection methods can be classified into three main categories. The first category focuses on the detection of the intersection or, in other words, the geometrically convergent points of the main vessel arcs. The second category benefits from the brightness and color features of the OD. The third category proposes an OD template and then the retinal region which resembles the template most is considered as the OD region [4]. In addition to these methods, there are also a vast number of OD detection algorithms. Some of these recent state-of-the-art methods are reported comparatively in table 2.1.

Fan *et al* proposed a structured learning based classifier model. The clipped OD region was used for training the classifier and the generation of the OD edge map, which was then thresholded and the OD patterns were located by applying the circular Hough transform (CHT). The test results for the DRIONS, MESSIDOR, and ONHSD retinal archives which have 110, 1200, and 99 images, respectively, were promising. However, the OD cases with severe peripapillary atrophy could not be located in this study. Additionally, there is a need for a vast number of retinal images to obtain a robust model [10].

Table 2.1. A comparison table for recent state-of-the-art studies. (Reproduced with permission from [1]. Copyright 2018 Elsevier.)

Author	Retinal archives	Success rate	OD overlap ratio	Execution duration (s)
Fan *et al* [10]	MESSIDOR	0.97	0.86	1.74
	DRIONS	0.98	0.84	
	ONHSD	0.99	0.83	
Bander *et al* [14]	MESSIDOR	0.97		0.007
	KAGGLE	0.967		
Tan *et al* [15]	DRIVE	0.8790	0.621	
Díaz-Pernil *et al* [16]	DRIVE	0.975	0.833	7.6
	DIARETDB1	0.977	0.843	16.3
Alshayeji *et al* [17]	DRIVE	1		0.4
	DIARETDB1	0.977		
	DMED	0.929		
	STARE	0.95		
Dai *et al* [18]	MESSIDOR			0.91
	DRIONS			0.908
	ONHSD			

Al-Bander *et al* came up with a sequential multiscale convolutional neural network (CNN). The KAGGLE and MESSIDOR retinal archives which have 10000 and 1200 images were used for training and testing the proposed system. This method had satisfactory results and required a very short execution time for testing retinal images. However, a variety and great number of data were still required for this algorithm, as for all the other supervised methods, in order to obtain a generalizable model which could also handle outlier anomalies [14].

Tan *et al* proposed a method which detected not only OD but also fovea and retinal vessels via a seven layered CNN. The CNN, which had four different outputs as OD, blood vessels, fovea, and background, was fed with three input channels extracted from the neighborhood of every effective point on the normalized retinal images. However, this method was executed only on the DRIVE retinal image archive and the performance was not promising enough. In spite of the fact that this study does not seem to be robust enough towards anomalies such as peripapillary atrophy and exudates, it still should be taken into consideration since simultaneously locating different retinal anatomy landmarks is an important contribution [15].

Díaz-Pernil *et al* came up with a novel OD localization algorithm which was composed of three consecutive algorithms, namely the AGP-color segmentator, Hamadani's binarization, and Hough circle cloud. Even if the performance of this method, which was tested on 129 images of DRIVE and DIARETDB1, was encouraging, the size and variability of these retinal archives were not enough to obtain a sufficiently robust model and was not tested on a retinal archive such as STARE which has various types of anomalies. However, the proposed method is still valuable since its image processing pipeline is novel and the algorithm was implemented for GPU architecture [16].

Alshayeji *et al* proposed an OD detection method whose main phase was established on an edge detection method which was influenced by the gravitational law. The tests of this method were performed on the DMED, DIARETDB1, and DRIVE retinal archives which have 169, 89, and 40 images, respectively. They also tested on only 20 images of the STARE retinal archive, which was originally composed of 81 images. The consistent performance rates across each of the four retinal archives were promising in terms of robustness and additionally the execution duration of the algorithm was low. However, they did not test the remaining images of STARE which have different sorts of anomalies and lesions [17].

Dai *et al* came up with a novel OD segmentation method which proposed a variational model with three energies. First, the OD was localized via a sparse coding based algorithm and the feature vectors were extracted by the Dense HueSIFT, Dense SIFT, and HoG algorithms. Then, the OD was segmented by evolving the curves driven by a variational model which is constituted of three energy terms towards the OD contours. The proposed method was tested on the ONHSD, MESSIDOR, and DRIONS retinal archives which have 99, 1200, and 110 images, respectively. The performance of this method was very satisfactory and the math behind the variational model was theoretically strong. However, the performance of the OD localization phase and the execution duration of the method were not demonstrated definitely in the article [18].

The proper localization and segmentation of the OD is still a difficult area of research because of lesions, uneven light distribution, noise, and abnormalities in retinal images [19]. The shape, color, and brightness derived properties and template matching based OD detection algorithms assume that the analyzed OD is always yellowish, bright, and regularly shaped. These sorts of algorithms have very satisfactory results when there are no anomalies or lesions. More and more, some of these algorithms may require long execution durations [6, 20, 21]. On the other hand, vessel convergence based algorithms exploit the structure and direction of the vessels to border the region of interest for the OD. These sorts of algorithms have promising results in suppressing the lesions whose shape or color are similar to the OD, but their proposed vessel convergence algorithms mostly require long execution durations [19, 22, 23]. Last but not least, the DL based OD detection algorithms have very good performance rates, but they also inherit the supervised learning based limitations, such as being biased on the training set and the need for a vast amount of data. As an example, a retinal archive training set which mostly contains diabetic macular edema (DME) will bias the representation of the DL system and make it learn to associate the temporal location of the OD and the existence of DME [24].

In this study, the proposed method contains two main phases. The first phase detects the convergence point of the main vessel arcs roughly via an enhanced version of a method which was been offered in a conference paper [25]. The enhancement of the first phase is provided by developing lesion removal and adaptive vessel thresholding algorithms. The second phase of this study mainly contains a fuzzy logic controller, novel feature extraction algorithms, circular pattern estimation, and an iterative approach applied to the whole algorithm in the case of outlier images by altering the initial parameters and color space.

2.2 Materials and methods

2.2.1 Materials

The DRIVE (Digital Retinal Images for Vessel Extraction) retinal archive was created mainly for retinal vessel detection research. It contains 40 retinal images and seven of them have symptoms of mild DR. The images were captured at a 565 × 584 pixel resolution and a field of view (FOV) of 45°. A mask corresponding to the FOV of each retinal image was provided. One half of the archive is used as a training set whereas the other half is used as a test set. The ground truth coordinates for the OD contours and centers were annotated by Professor Anna L Ells, an ophthalmologist specialized in retina at the Alberta Children's Hospital Research Institute, Calgary University, Canada, and provided by Professor Emeritus Rangaraj M Rangayyan at the Department of Electrical and Computer Engineering, Calgary University, Canada [26–29].

The STARE (Structured Analysis of the Retina) retinal archive contains 397 images which are of 700 × 605 pixel resolution and have an FOV of 35°. Each image is clipped at the top and bottom of the FOV and diagnosed as having at least one of the 13 different retinal anomalies. The ground truth coordinates of the OD center locations for 81 images are given [30, 31].

DIARETDB1 is a retinal archive which is proposed for DR related research. This retinal archive is composed of 89 images where 84 of the images are identified as having mild non-proliferative or more severe signs of DR. Each image is captured with a 1500 × 1152 pixel resolution and an FOV of 50° with changing camera settings. The ground truth coordinates of different lesions were annotated by four expert ophthalmologists but no ground truth was provided for OD centers or contours. The ground truth locations of the OD centers of the DIARETDB1 retinal images were annotated by the expert ophthalmologist Dr Ömer Takeş [32, 33].

DIARETDB0 is a retinal archive which is proposed for DR related research. This retinal archive is composed of 130 images where 110 of the images were identified as having mild non-proliferative or more severe signs of DR. Each image was captured with a 1500 × 1152 pixel resolution and an FOV of 50° with unknown imaging settings. The ground truth locations of the OD centers of the DIARETDB0 retinal images were annotated by the expert ophthalmologist Dr Ömer Takeş [34, 35].

2.2.2 Methods

The flowchart of the algorithms of this study is sketched in figure 2.1 [1]. The algorithms, such as 'estimation of the convergence point of vessels', 'preprocessing', 'circular pattern detection', and 'feature extraction' based on 'yellowish threshold estimation' were proposed previously [12, 17, 18]. The sub-algorithms of 'preprocessing', such as 'lesion removal', 'light equalization', and 'vessel enhancement' are borrowed from previous articles [25, 36–40]. However, the components such as

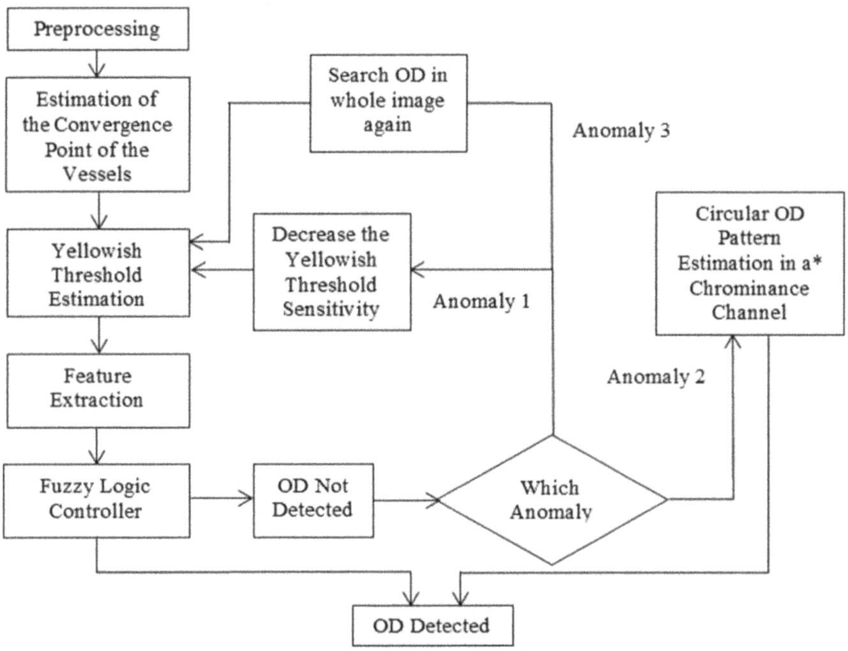

Figure 2.1. The flow diagram of the proposed algorithm. Reproduced with permission from [1]. Copyright 2018 Elsevier.

algorithm 2.1 and 'adaptive binary thresholding' are fully novel. 'Estimation of the convergence point of vessels', which is an enhanced and robust version of a previously proposed primitive method, contains algorithms 2.2 and 2.3 as extra novelties [25]. The first two 'feature extraction' algorithms were proposed in previous articles but the third one is novel and acts as an effective biomarker for OD center detection [36].

2.2.2.1 Preprocessing

Preprocessing contains some important sub-algorithms such as 'light equalization', 'vessel enhancement', 'adaptive binary thresholding', and 'lesion removal'.

2.2.2.1.1 Light equalization

Uneven light distribution is a common and basic problem in the retinal image analysis field. In this study, contrast limited adaptive histogram equalization (CLAHE), which is a popular method in medical image analysis, is applied to the intensity channel of the HIS (hue, intensity, saturation) color space of the retinal images, as shown in figure 2.2 [1, 25, 36–38].

In the CLAHE method, the image is split into $i \times j$ frames which are generally selected as 8×8 frames and later the histogram ($h_{i,j}$) of each frame is obtained. Thereafter, a clip limit β is calculated, as described in equation (2.1), to be able to set a limit for the noisy intensity pixel values which will be obtained as a result of the latter contrast improvement steps. The $s_{\max}$ slope value can vary with respect to the image properties and the clip factor α can be set between 0 and 100. Afterwards, the extra grayscale pixel quantities which are higher than β are equally distributed to the remained histogram bins.

$$\beta = \frac{M}{N}\left(1 + \frac{\alpha}{100}(s_{\max} - 1)\right). \tag{2.1}$$

Later, the cumulative density function (CDF) ($f_{i,j}$) of the histogram of each frame is obtained as described in equation (2.2), M and N, respectively, represent the quantity of pixels and grayscales for each frame, where the range of n is set from 1 up to N−1:

$$f_{i,j}(n) = \frac{(N-1)}{M}\sum_{k=0}^{n} h_{ij}(k) n = 1,2,3, \ldots, \ N-1. \tag{2.2}$$

Lastly, the final values of the pixels are set by bilinear interpolation of the CDF of their nearest vertical and horizontal adjacent frames [38, 41].

2.2.2.1.2 Vessel enhancement

In the field of medical image analysis, the Frangi vesselness filter (FVF) is a useful method for detecting vessel patterns. The FVF exploits the geometric relationships between the eigenvalues of the Hessian matrix which is obtained by second order Gaussian derivatives of the images. The vessel patterns of an example retinal image which are emphasized by the FVF can be observed in figure 2.3 [1, 25, 39].

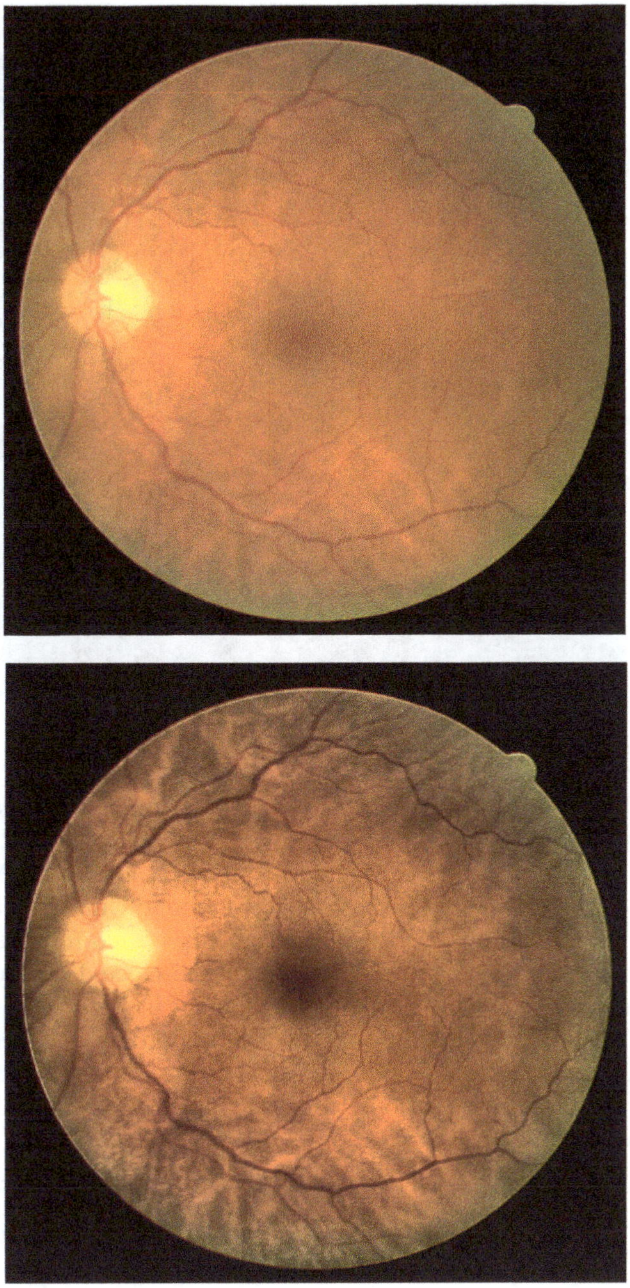

Figure 2.2. The before and after effect of CLAHE algorithm is shown on a) and b) images, respectively. Reproduced with permission from [1]. Copyright 2018 Elsevier.

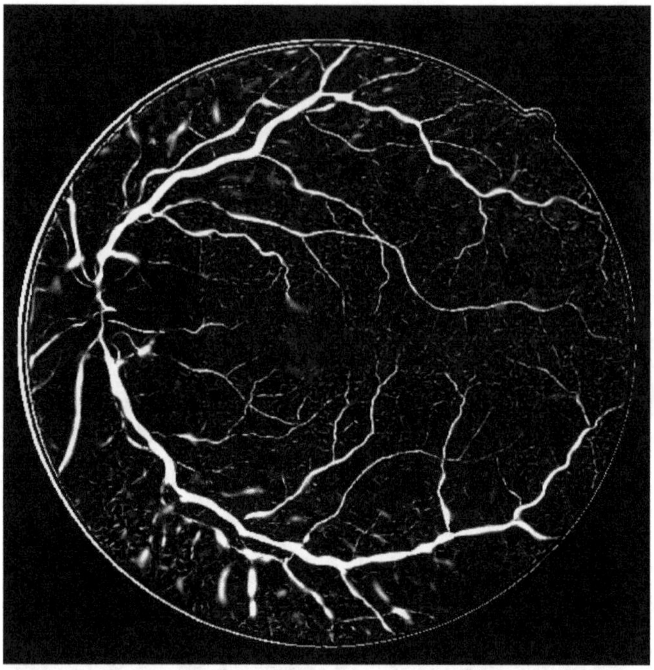

Figure 2.3. The emphasized vessel patterns of FVF response. Reproduced with permission from [1]. Copyright 2018 Elsevier.

2.2.2.1.3 Adaptive binary thresholding and lesion removal

As the last sub-algorithm of the preprocessing step, the artefacts or lesions of the retinal images are erased to a great extent in algorithm 2.1 [1]. First F, which represents the response of FVF, is thresholded by t_V. The obtained binary matrix result is set to B which also contains n connected components symbolized as S_n. The minor and major axis lengths are the measurements related to the ellipse which has the same normalized second central moments as a connected component. Solidity is the proportion of the pixels in the region to the ones in the convex hull. The minor axis, major axis, surrounding ellipse, and lesion region are depicted in blue, green, red, and white, respectively, in figure 2.4(c) [1]. It is empirically realized that the connected components whose ratio between the major and minor axis lengths is less than 5 and whose solidity value is more than 0.3 are likely to be detected as lesions [40]. Thus, the connected components, S_i, which are detected as lesions, are erased in the matrix B. Thereafter, the pixels in B are morphologically thickened and the connected components larger than 200 pixels are removed from B. Then, the pixels in B are morphologically skeletonized and thickened, after which B is reshaped as a square matrix by adding redundant zero pixels along the shorter axis in order to be able to apply the following vessel convergence algorithm. Finally, if the total number of the pixels in the matrix B is below t_c then the threshold t_v is decreased by 0.001 and all the algorithm is executed again, iteratively, starting from the beginning. Otherwise, the algorithm halts the iterative execution process by yielding the last version of the binary matrix B.

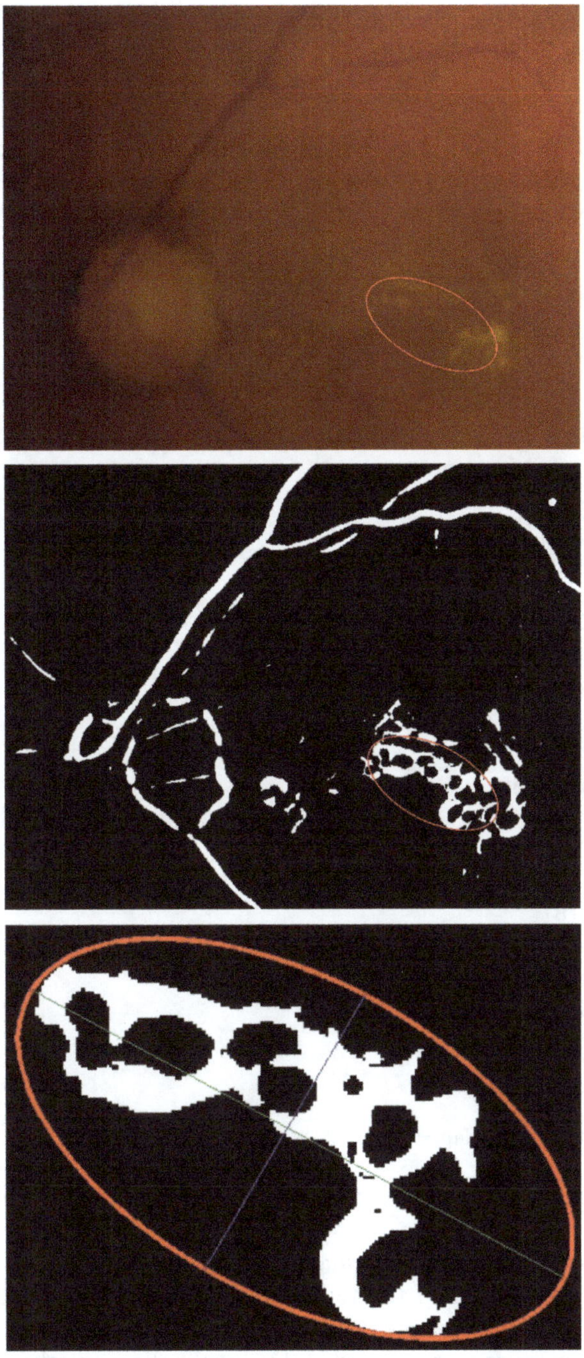

(a) (b) (c)

Figure 2.4. The lesion elimination process (a) the original image having the hard exudate lesion encircled in the red ellipse (b) the FVF response of the original image whose lesion is highlighted with the red ellipse (c) zooming into the red ellipse which encapsulates the lesion with its major and minor axes. Reproduced with permission from [1]. Copyright 2018 Elsevier.

Algorithm 2.1. Adaptive binary thresholding and lesion removal algorithm. (Reproduced with permission from [1]. Copyright 2018 Elsevier.)

Initialization:
$F_{initial} \leftarrow F$, $B \leftarrow 0$ **// F: response of FVF, B: thresholded binary vesselness image**
$t_c \leftarrow 5000$, $t_v \leftarrow 0{:}01$ **//tc: threshold for number of pixels in B, tv: threshold used over F to get B**
total_pixel ← 0, visit_flag ← 0

begin
repeat
$\forall p_{ij} \in F$, $F(p_{ij}) > t_v$, $B(p_{ij}) \leftarrow 1$
$a \neq b$, $a \leqslant n$, $b \leqslant n$, $1 \leqslant n$, $1 \leqslant i \leqslant n$
$\forall q_{ij} \in B$, $B(q_{ij}) \leftarrow 1$, $q_{ij} \in S_a$, $q_{ij} \not\subset S_b$, $S_i \subset B$
//S_a, S_b, S_i: each single different connected component representing vessel patterns
for each S_i do
Solidity $\leftarrow \frac{(\text{Region of } S_i)}{(\text{Convex Hull of } S_i)}$
AxisLengthRatio $\leftarrow \frac{(\text{Major Axis Length of } S_i)}{(\text{Minor Axis Length of } S_i)}$
if Solidity >0.3 and AxisLengthRatio <5 **then**
$\forall q_{ij} \in S_i$, $B(q_{ij}) \leftarrow 0$
end if
end for
Thicken(B)
for each S_i do
$\forall q_{ij} \in S_i$, $B(q_{ij}) < 200$, $\sum B(q_{ij}) \leftarrow 0$
end for
Skeletonize(B)
Thicken(B)
MakeSquareMatrix(B)
visit_flag ← 1
$\forall q_{ij} \in B$, total_pixel $\leftarrow \sum B(q_{ij})$
if total_pixel $< t_c$ **then**
$t_v \leftarrow t_v - 0.001$
$F_{initial} \leftarrow F$
end if
until total_pixel $< t_c$ or visit_flag = 0
end

2.2.2.2 Estimation of the convergence point of the vessels

The layout and orientation of the retinal blood vessels have been utilized for OD region localization by some previous studies [25, 26]. In this study, a formerly proposed vessel convergence based OD localization method is improved by developing some extra enhancements and a novel valley regularization algorithm as in algorithms 2.2 and 2.3 [1, 25]. In algorithm 2.2 the w_size parameter, which is the edge length of a square window W_{ab}, is empirically set as 20, 20, 20, and 16 for the DIARETDB1, DIARETDB0, STARE, and DRIVE retinal archives, respectively. The binary matrix B is divided into these square windows and the number of pixels corresponding to each W_{ab} window is set to the (a,b) coordinate of the vesselPixelCounts matrix. Thereafter, to the vesselPixelCounts matrix is applied a 2D Gaussian smoothing kernel with a standard deviation of 1 and a mean of 4. The selection reason for these two parameters of the kernel is the fact that an average OD roughly fits to four adjacent tiles of the vesselPixelCounts matrix.

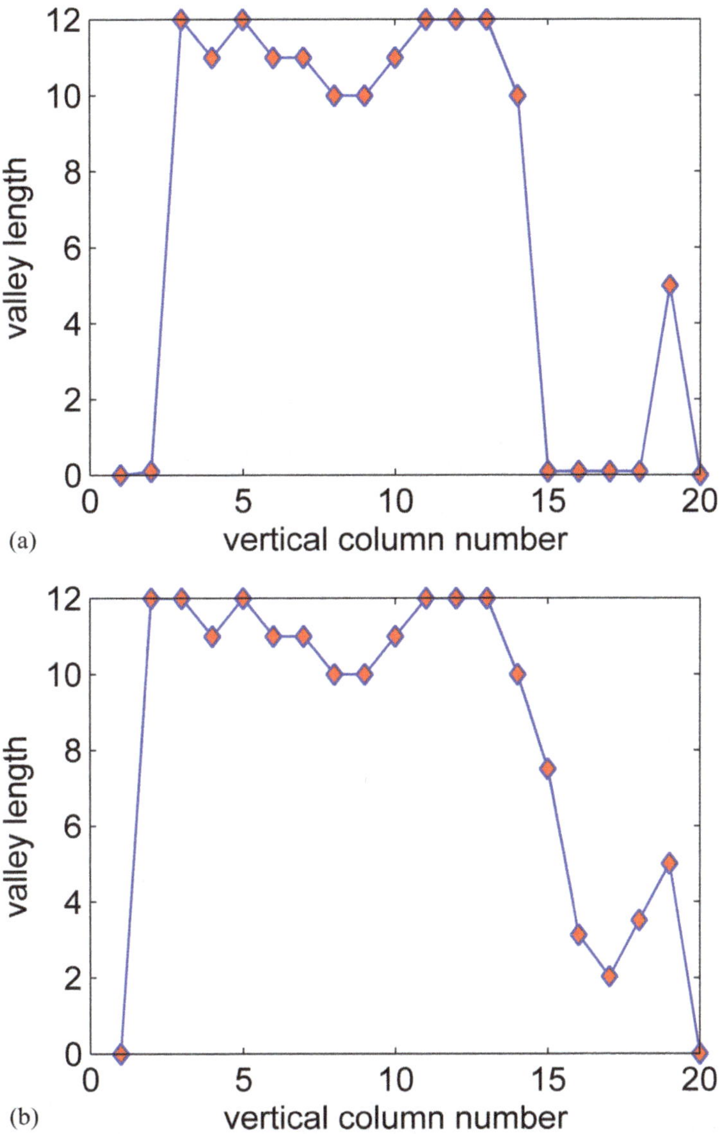

Figure 2.5. The before and after effect of valley length regularization algorithm is shown on a) and b) images, respectively. Reproduced with permission from [1]. Copyright 2018 Elsevier.

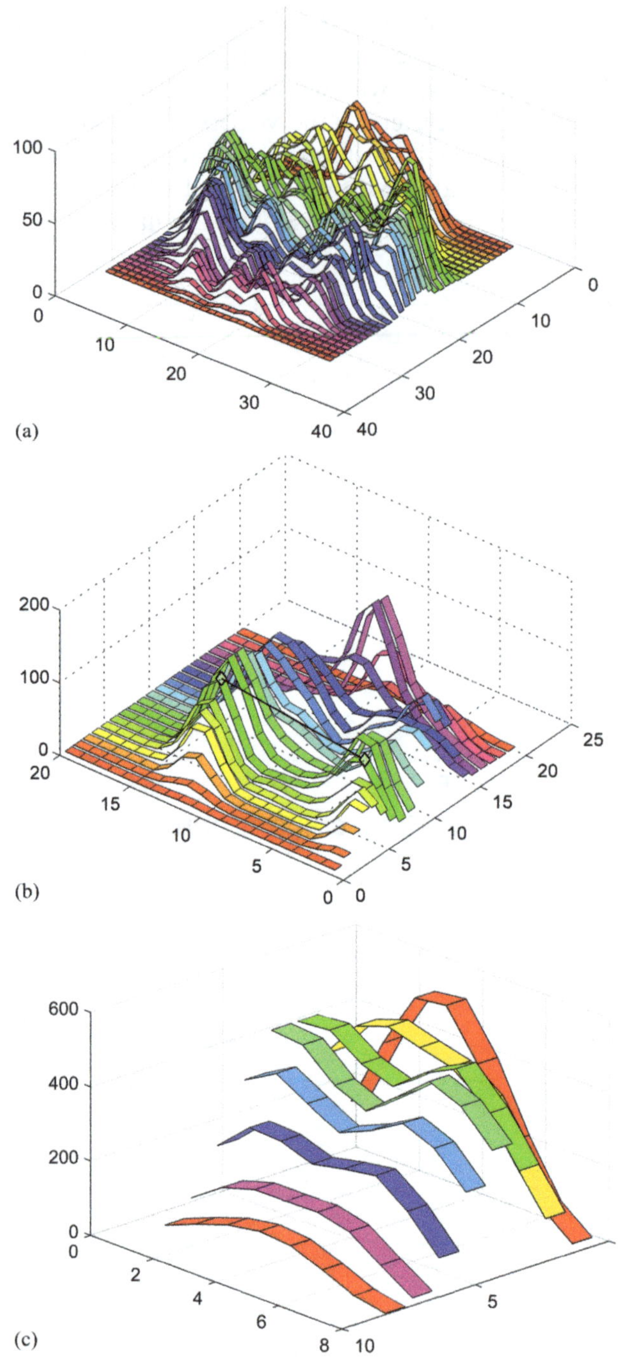

Figure 2.6. The vessel pixel distributions for 32x32, 16x16 and 8x8 windows size values are visualized on a), b) and c) images, respectively. Reproduced with permission from [1]. Copyright 2018 Elsevier.

The w_size parameter for the DRIVE dataset is decided after a few experiments, as can be observed in figure 2.6 [1]. The intricacy as well as the number of the valleys and hills become too much if the w_size is set as 32, as can be seen in figure 2.6(a) [1]. In contrast, if w_size is set as 8 then the hills and valleys become much less complex but also the resolution of the OD location decreases, as shown in figure 2.6(c) [1]. Last but not least, when w_size is empirically set as 16 then it is observed that the resolution of the OD location is reasonable and single vertical valleys are present, obtained as shown in figure 2.6(b) [1].

Afterwards, each column of the vesselPixelCounts matrix is copied to verticalCrossSection vectors, as can be observed in the form of vertical ribbons in figure 2.7(b) [1]. The local maxima of the verticalCrossSection vectors whose values are more than 10 and symbolized as q_j are detected and the maximums of these local maxima are set as profilePeaks for each specific column. If a verticalCrossSection vector has more than two local peaks then the distance between the lowest and highest peaks are set as the valleyLengths, seen as the black line in figure 2.7(b), whereas the minimum value restricted on that valley region is set as valleybottomValues [1]. Algorithm 2.2 also handles the cases where there is at most one peak per verticalCrossSection vector. Thereafter, the cvlmv, cvbv, cpp, and cpp2 candidate verticalCrossSection vector lists are privileged as the final candidate column vectors, since the fact that the pixel density patterns of the vesselPixelCounts geometrically converge to a relatively higher intersection point which is the OD location itself, as can be observed in figure 2.7(a) [1]. In other words, the closer one gets to the OD, the larger the valleybottomValues and profilePeaks and conversely the shorter the valleyLengths observed. Finally, the column which has the maximum profilePeaks to valleyLengths ratio among candidate columns is set as the estimatedColumn whereas the estimatedRow is calculated as the row which has the highest value on the estimatedColumn, as can be seen in figure 2.7(c) [1].

Algorithm 2.3 proposes a regularization solution for the robustness of algorithm 2.2 by handling the cases where at least one of the peaks on a horizontal vessel arc does not have any corresponding peak on the other horizontal arc. This kind of absence of corresponding peaks causes some types of error during candidate and ratio calculations. In algorithm 2.3 the dip points on the valleyLengths which have a value of 0.1 are represented as $_{pi}$, whereas p_r and p_k, respectively, represent the nearest right-side non-dip point and the points between $_{pi}$ and p_r. Similarly, p_l and p_j, respectively, represent the nearest left side non-dip point and the points between $_{pi}$ and p_l. Finally, the $_{pi}$ values are updated using the p_l, p_r, p_j, and p_k values with respect its membership to cpp. The effect of regularization in algorithm 2.3 over valleyLengths vectors can be observed in figure 2.5 [1].

Algorithm 2.2. Vessel convergence point estimation algorithm. (Reproduced with permission from [1]. Copyright 2018 Elsevier.)

Initialization:
define vesselPixelCounts matrix(w_size,w_size)
define profilePeaks, valleyLengths, valleybottomValues
matrix(w_size, 1)

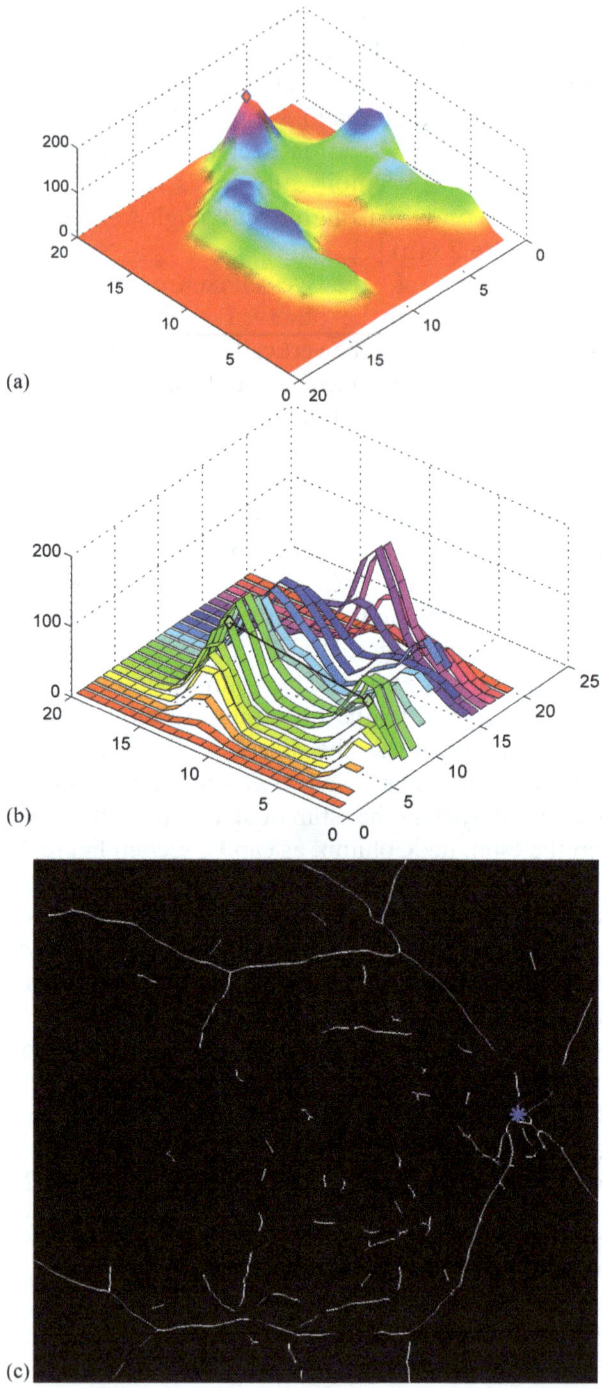

Figure 2.7. The estimation process of OD neighborhood a) the vesselPixelCounts matrix is visualized as a surface and OD is marked by red diamond b) the vertical profile of vesselPixelCounts matrix is visualized via ribbons c) the final estimated OD point is shown on the vessel skeleton via a blue star mark. Reproduced with permission from [1]. Copyright 2018 Elsevier.

```
define W_ab matrix // W_ab: the matrix which holds the vessel distribution values
1⩽a⩽w_size, 1⩽b⩽w_size

begin
for each W_ab do
∀p_ij ∈ W_ab, vesselPixelCounts(a, b) = ∑B(p_ij)
end for
vesselPixelCounts ← GaussianSoften(vesselPixelCounts, 4, 1)
for each i of w_size do
verticalCrossSection ← ith column of vesselPixelCounts
temp_pks ← local maxima of verticalCrossSection
∀q_j ∈ temp_pks, q_j > 10, profilePeaks(i) ← max(q_j)
if card{q_j} ⩾2 then //card{qj}: cardinality/count of the peaks of vertical cross section
valley ← verticalCrossSection(min(q_j):max(q_j))
valleyLengths(i) ← card{valley} //card{qj}: cardinality/count of the peaks of vertical cross
section
if min(valley) ⩾0 then
valleybottomValues(i) ← min(valley)
else
valleybottomValues(i) ← 0.1
end if
else if card{q_j} = 1 then
valleyLengths(i) ← 0.1
if max(q_j) >0 then
valleybottomValues(i) ← max(q_j)
else
valleybottomValues(i) ← 0.1
end if
else
valleyLengths(i) ← max(valleyLengths)
valleybottomValues(i) ← 0.1
end if
end for
cpp ← largest first 5 local maxima of profilePeaks
cpp2 ← above mean of profilePeaks
valleyLengths ← valley_length_regularizer(valleyLengths, cpp)
cvlm ← largest first 5 local minima of valleyLengths
cvbv ← largest first 5 local minima of valleybottomValues
candidates ← ((cpp or cvlm) and cpp2) or cvbv
ratios ← profilePeaks(Candidates) / valleyLengths(Candidates)
estimatedColumn ← the column of max(ratios)
estimatedRow ← the row of max(estimatedColumn)
end
```

Algorithm 2.3. Valley length regularization algorithm. (Reproduced with permission from [1]. Copyright 2018 Elsevier.)

Initialization:
$\forall$pi $\in$ valleyLengths, valleyLengths ($_{pi}$) $\leftarrow$ 0.1
$\forall p_l \in$ valleyLengths, valleyLengths (p_l) > 0.1, 0 < l < i
$\forall p_r \in$ valleyLengths, valleyLengths (p_r) > 0.1, 0 < r < i
valleyLengths (p_j) $\leftarrow$ 0.1, l < j < i, 0 $\le$ j
valleyLengths(p_k) $\leftarrow$ 0.1, i < k < r, k $\le$ w_size
i $\leftarrow$ 1

begin
for each pi do
if $\forall p_l = 0$ **and** $\forall p_r \neq 0$ **then**
valleyLengths(pi) $\leftarrow \frac{\text{valleyLengths}(p_l)}{(i - \min(j))}$
else if $\forall p_l \neq 0$ **and** $\forall p_r = 0$ **then**
valleyLengths(pi) $\leftarrow \frac{\text{valleyLengths}(p_r)}{(i - \max(k))}$
else if $\forall p_l \neq 0$ **and** $\forall p_r \neq 0$ **then**
if $\in$ cpp **then**
valleyLengths(pi) $\leftarrow \left(\frac{\text{valleyLengths}(p_l)}{4}\right) + \left(\frac{\text{valleyLengths}(p_r)}{4}\right)$
else
valleyLengths(pi) $\leftarrow \left(\frac{\text{valleyLengths}(p_l)}{2}\right) + \left(\frac{\text{valleyLengths}(p_r)}{2}\right)$
end if
else
valleyLengths(pi) $\leftarrow$ w_size
end if
i $\leftarrow$ i + 1
end for
end

2.2.2.3 Yellowish threshold estimation

In this study, a yellowish threshold value over the green channel histogram is estimated iteratively as in algorithm 2.4 for the latter two feature extraction algorithms [1, 36, 37]. The proportion of radii_max which is an average radii for a regular OD region to the whole masked image is represented as t_r. The radii_max value is, respectively, set as 130, 130, 50, and 75 for the DIARETDB1, DIARETDB0, DRIVE, and STARE retinal archives. In this algorithm, the yellowish threshold value is estimated as the point where the ratio of the remaining regions of the right and left sides is approximately t_r and an example iteration process is depicted in figure 2.8 [1]. In algorithm 2.4, the I represents the green channel of the retinal image and card{I} shows the count of the pixels in I. L_{Low} shows the lowest value in the histogram whereas the L_{High} is the highest one. L_{Temp} is the temporary threshold estimation and iteratively shifted to the left until the ratio of the region of its right side to the whole histogram converges close enough and settles down to a balanced estimation point which is represented by a yellow vertical line.

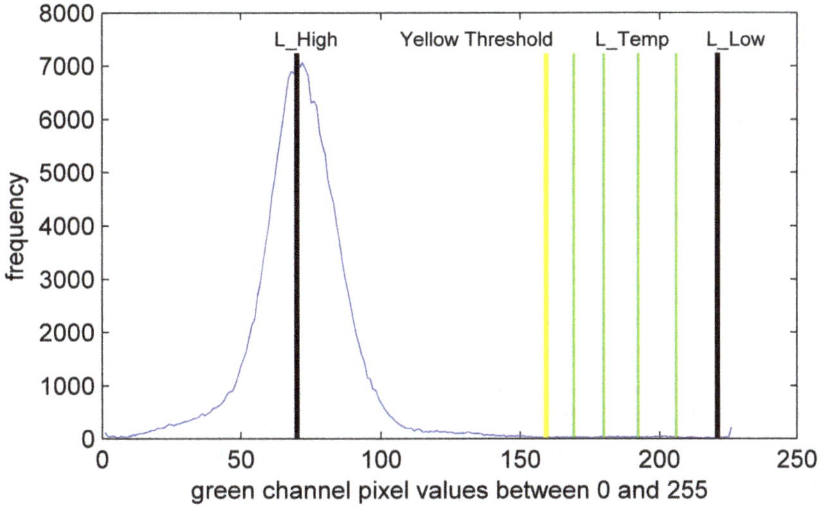

Figure 2.8. The iterative yellow threshold estimation process. Reproduced with permission from [1]. Copyright 2018 Elsevier.

Algorithm 2.4. Yellow threshold estimation algorithm. (Reproduced with permission from [1]. Copyright 2018 Elsevier.)

Initialization:
$H \leftarrow$ histogram of I
radii_max $\leftarrow 50$
$t_r \leftarrow \frac{(\text{radii\_max radii\_max } \pi)}{\text{card}\{I\}}$
$L_{High} \leftarrow \max(H(i)),\ L_{HighIndex} \leftarrow i,\ 1 \leqslant i \leqslant 255$
$L_{Low} \leftarrow \max(H(i)),\ L_{LowIndex} \leftarrow i,\ 1 \leqslant i \leqslant 255$

begin
repeat
$L_{TempIndex} \leftarrow \left(L_{LowIndex} - \frac{(L_{LowIndex} - L_{HighIndex})}{10}\right)$
$R \leftarrow j,\ L_{TempIndex} < j < L_{LowIndex}$
ratio $\leftarrow \frac{\text{card}\{R\}}{\text{card}\{I\}}$
until ratio $\geqslant t_r$
Yellow Threshold $\leftarrow L_{TempIndex}$
end

2.2.2.4 Circular OD pattern estimation

The first phase of the study passes the estimated location of the convergent point of the vessels to the second phase as the center of the rectangular region which is clipped from the morphologically closed form of the whole retinal image for a more focused searching of the OD pattern. The size of the clipped region is set as 350 × 350, 350 × 350, 240 × 240, and 200 × 200 for the DIARETDB1, DIARETDB0, DRIVE, and STARE retinal archives, respectively. Thereafter, the canny edge detection (CED) algorithm is applied to the clipped rectangular image with an empirically decided threshold value of 0.1.

Afterwards, only the filtered candidate circular patterns which are in a reasonable minimum and maximum radius value range are detected via CHT, as in former studies [36, 37]. The radius range is set between 45 and 75 for STARE, 30 and 50 for DRIVE, and 70 and 130 for the DIARETDB1 and DIARETDB0 retinal archives. The estimated circular OD patterns are shown in figure 2.9 [1].

2.2.2.5 Feature extraction

In this study, three types of features are extracted from the estimated circular OD patterns. The first feature is extracted by multiplying the percentage and count of the green channel pixels whose values are larger than the yellowish threshold value [1, 36]. The first type of feature provides a useful insight because of the fact that the OD is usually yellowish and its robustness is not disturbed by the hard and soft exudate lesions since they are mostly erased by algorithm 2.1.

In order to obtain the second feature, the green channel pixels whose values are larger than the yellowish threshold values and shapes in the form of connected components larger than an experimentally determined threshold are filtered. Then, the second feature is calculated as the ratio of the region of the filtered components to the estimated circular OD pattern's region. The first feature is backed by the second feature by decreasing the effect of the random yellowish noise [1].

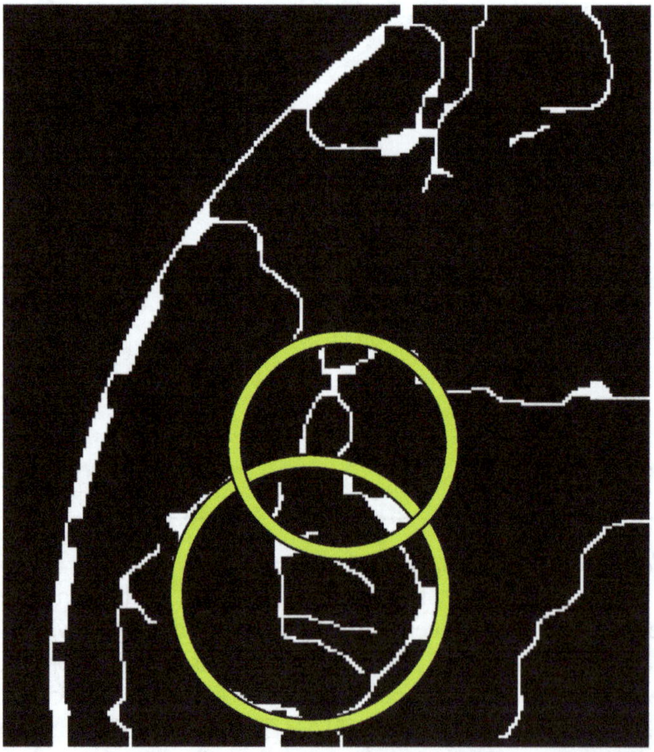

Figure 2.9. The OD estimation process on circular patterns. Reproduced with permission from [1]. Copyright 2018 Elsevier.

The third feature is extracted by the product of the number of branching points and the pixels on the retinal blood vessel skeleton, which is obtained as a side product of algorithm 2.1, as shown in figure 2.10 [1]. This novel feature is not only a very strong biomarker for the OD region but is also robust against light, color, and atrophic anomalies.

2.2.2.6 Fuzzy logic controller

A Mamdani type fuzzy logic controller (FLC) is proposed to classify the candidate circular OD patterns and their extracted features as 'OD', 'OD intersection', and 'not OD'. The design of the input membership functions are inspired by the probability distribution of all of the detected circular patterns in the DRIVE retinal archive. The output membership functions are shown in figure 2.11 whereas the normalized probability distributions and the corresponding designed input membership functions can be observed in figures 2.12 and 2.13, respectively [1]. It is worth noting that the membership functions of the proposed FLC are fully designed by using statistical information derived only from DRIVE, but is used not only on DRIVE but also on the DIARETDB0, DIARETDB1, and STARE retinal archives during tests.

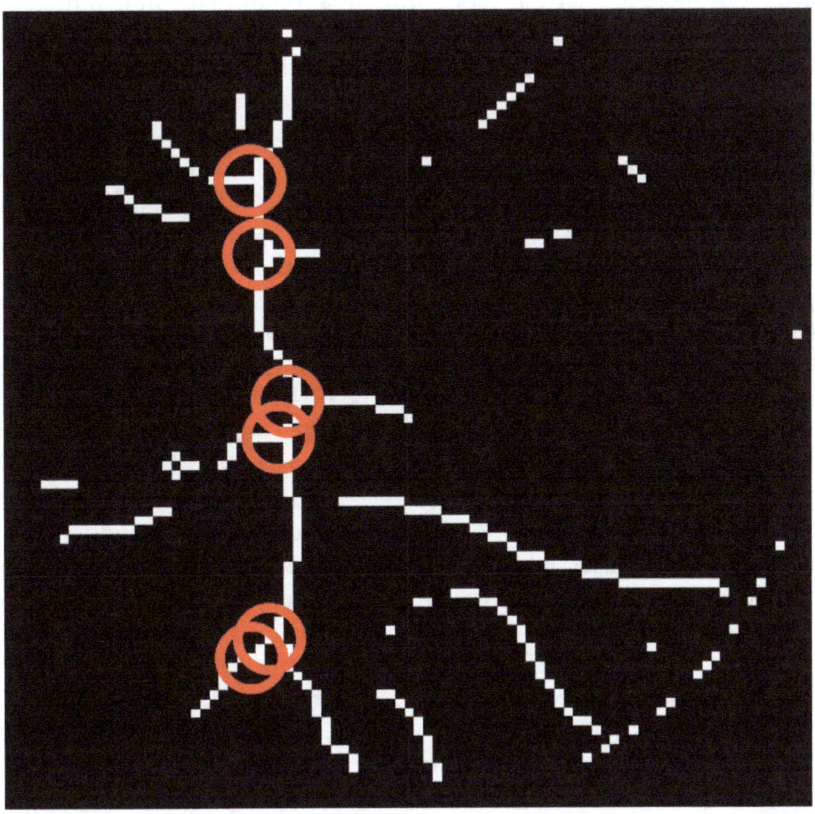

Figure 2.10. The branch points on vessel skeleton are marked by red circles. Reproduced with permission from [1]. Copyright 2018 Elsevier.

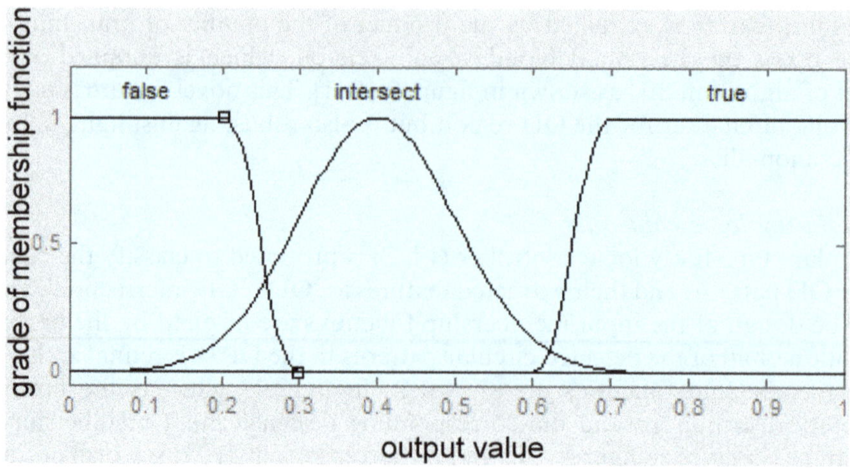

Figure 2.11. The output membership function. Reproduced with permission from [1]. Copyright 2018 Elsevier.

2.2.2.7 Anomaly detection

It is realized that there are certain types of main anomalies for non-standard images in the DRIVE retinal archive. The anomaly definition is based on the cases when there are no sensible circular OD patterns or the FLC does not predict any of the candidate circular patterns as the OD.

Anomaly type 1 is depicted as the case when the estimated yellowish threshold is too high since the contrast is quite low and the energy of the image is conversely very high. The high yellowish threshold value causes the first and second features to be very low or even zero while the third feature is not affected by this anomaly and is still high enough. In order to overcome this anomaly type, the sensitivity parameter of the yellowish threshold is decreased and the rest of the steps are iteratively executed until the estimated value of the yellowish threshold settles on a reasonable value so that the FLC predicts at least one candidate circular pattern as the OD. An example retinal image depicting the anomaly type 1 case can be observed in figure 2.14(a) [1].

Anomaly type 2 describes the case when all the features are very low which indicates that there is no apparent circular OD pattern in the grayscale or green channel, as can be seen in figure 2.14(b) [1]. Hence, the original RGB image is converted to the $L^*a^*b^*$ (luminosity, a^* chrominance, b^* chrominance) channel due to the fact that it is observed that the a^* chrominance channel contains very discriminative information on yellowish regions. Afterwards, the OTSU algorithm is applied to reduce the histogram down to four color tones, and then morphological closing, opening, as well as CED operations are applied to that color channel [42]. Thereafter, the strongest three circular OD patterns are detected by the CHT and some extra candidate circles are added when the detected circles are close to each other. The reason of adding extra candidate circles is the fact that even though the a^* chrominance channel provides much better OD contour patterns it is not enough to precisely emphasize real OD contours. Finally, the candidate circular pattern whose third feature is the highest one is decided to be the searched OD circular pattern.

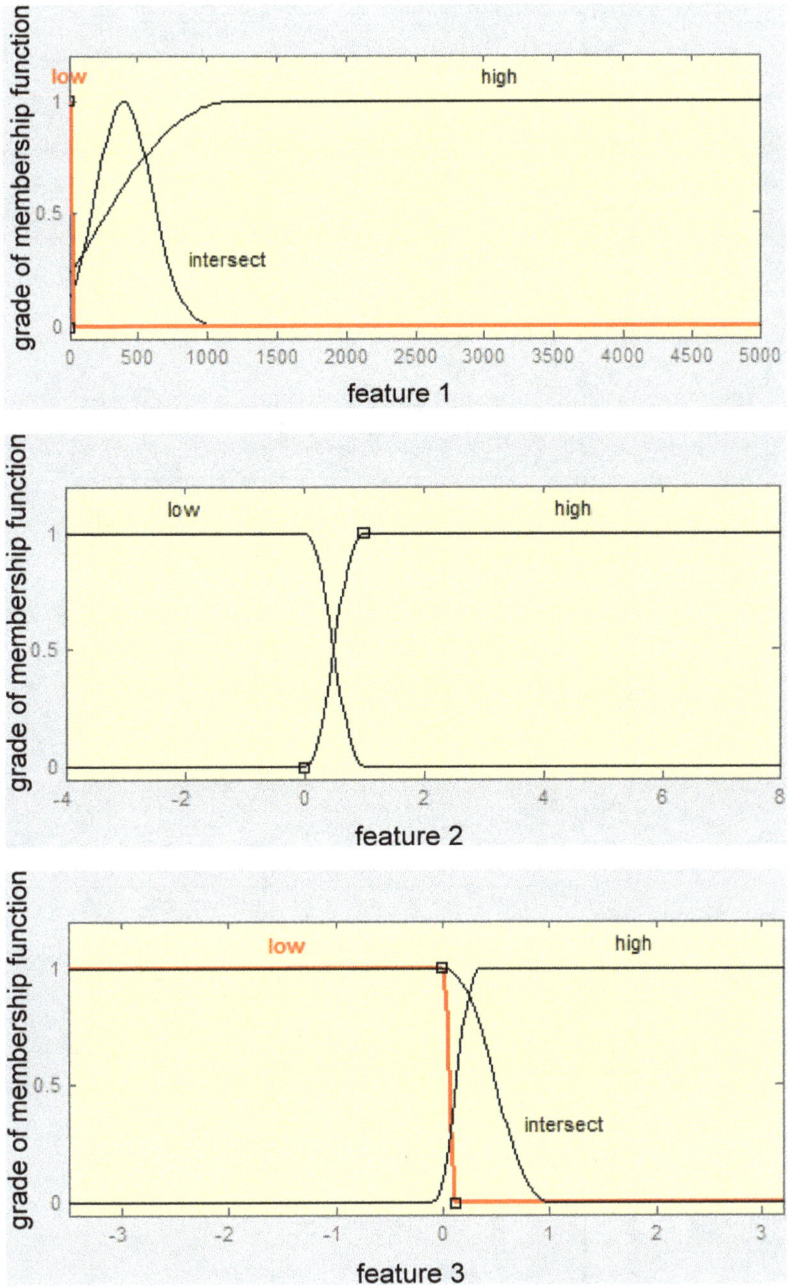

Figure 2.12. The membership functions of the extracted three features are shown on a), b) and c) images, respectively. Reproduced with permission from [1]. Copyright 2018 Elsevier.

If the first two anomaly types are not detected while the highest third feature among the candidate OD circles is below 400, which is a very low value with respect to the third feature of regular OD cases, then this sort of situation is diagnosed as

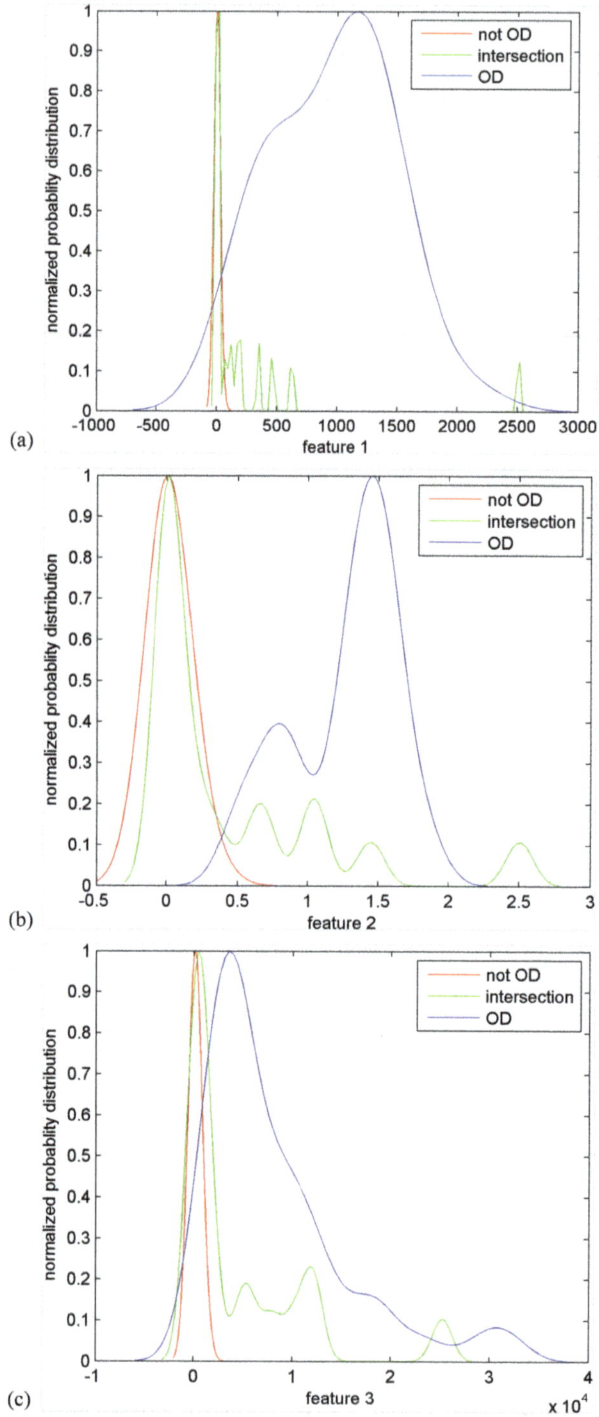

Figure 2.13. The probability distributions of the extracted three features are shown on a), b) and c) images, respectively. Reproduced with permission from [1]. Copyright 2018 Elsevier.

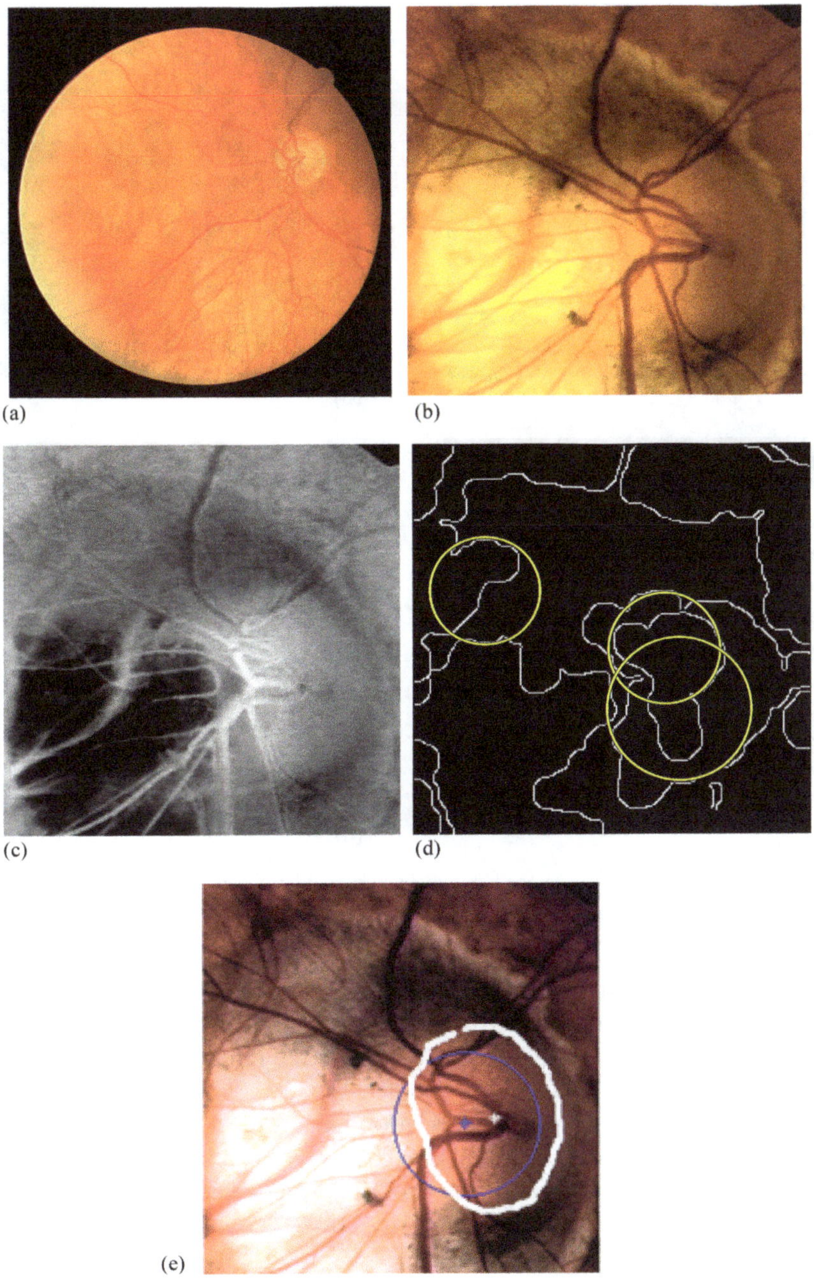

(a) (b) (c) (d) (e)

Figure 2.14. An anomaly type 1 image and OD detection steps on an anomaly type 2 image a) anomaly type 1 image b) anomaly type 2 image c) a* chrominance channel of anomaly type 2 image d) the visualization of the detected circular patterns after applying OTSU e) The final estimated encompassing blue circle and the ground truth counters marked as white.Reproduced with permission from [1]. Copyright 2018 Elsevier.

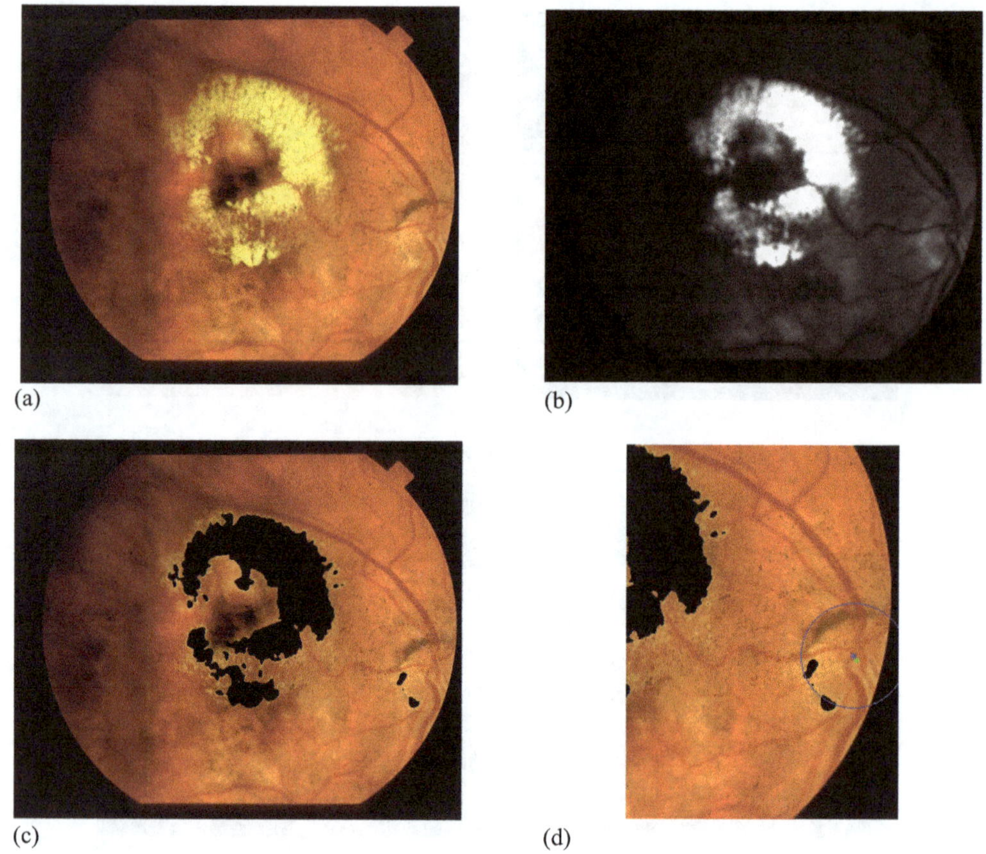

Figure 2.15. The filtering steps for an image having widely spread lesions in STARE a) the original image b) the response of the filter c) the filtered image d) the OD detection results by a blue circle with a green star on its center. Reproduced with permission from [1]. Copyright 2018 Elsevier.

anomaly type 3 [1]. The main reason for this anomaly type is the failure of the vessel convergence algorithm and in order to get rid of this problem the previously applied vessel convergence algorithm is blocked and then the circular OD pattern search is initiated in the whole image again.

Lastly, a simple and novel filter is proposed for the images which contain prevalent soft and hard exudate lesions, particularly in the STARE retinal archive. The highest 10% of the histogram of the product of the green and hue color spaces of this kind of retinal images is erased, as can be observed in figure 2.15 [1, 40].

2.3 Evaluation and results

The performance evaluation of the proposed method is accomplished with respect to three criteria. The first one is the Euclidean distance between the centers of the ground truth and detected OD patterns. The second criterion is the overlap which is the ratio of the intersection and the union of the regions of the ground truth and detected OD patterns, as shown in equation (2.3) [1]. The third criterion is the execution duration of the proposed method. The OD contour coordinates are only provided for the DRIVE retinal archive while the OD center coordinates are obtained for all four retinal archives. Thus, the second evaluation criterion is analyzed only for DRIVE whereas the first and third evaluation criteria are inspected for all four retinal archives. The statistics on the performance of all the evaluation criteria can be found in tables 2.2, 2.3, and 2.4 [1].

$$\text{Overlap} = \frac{(\text{Annotated Region} \cap \text{Detected Region})}{(\text{Annotated Region} \cup \text{Detected Region})}. \tag{2.3}$$

The OD center detection rates are 91.35% for STARE, 90% for DIARETDB0, and 100% for DIARETDB1 and DRIVE. In this study, the success criteria for the

Table 2.2. The statistics for the outcomes of the DRIVE retinal archive. (Reproduced with permission from [1]. Copyright 2018 Elsevier.)

Criterion	OD center distance error (pixels)	OD overlap ratio	Execution duration (s)
Average	6.6	0.778	13.96
Min	1.02	0.373	11.58
Max	33.15	0.917	53.92
SD	6.57	0.133	7.8

Table 2.3. The statistics for the outcomes of the DIARETDB0 and DIARETDB1 retinal archives. (Reproduced with permission from [1]. Copyright 2018 Elsevier.)

Criterion	OD center distance error (pixels) DIARETDB0	Execution duration (s)	OD center distance error (pixels) DIARETDB1	Execution duration (s)
Average	16.74	32.07	11.75	25.26
Min	1.43	18.95	1.12	18.23
Max	77.78	121.07	103.5	110.15
SD	14.68	22.38	15.8	13.73

Table 2.4. The statistics for the outcomes of the STARE retinal archive. (Reproduced with permission from [1]. Copyright 2018 Elsevier.)

Criterion	OD center distance error (pixels)	Execution duration (s)
Average	11.82	24.58
Min	1	7.49
Max	55	68.69
SD	12.29	16.98

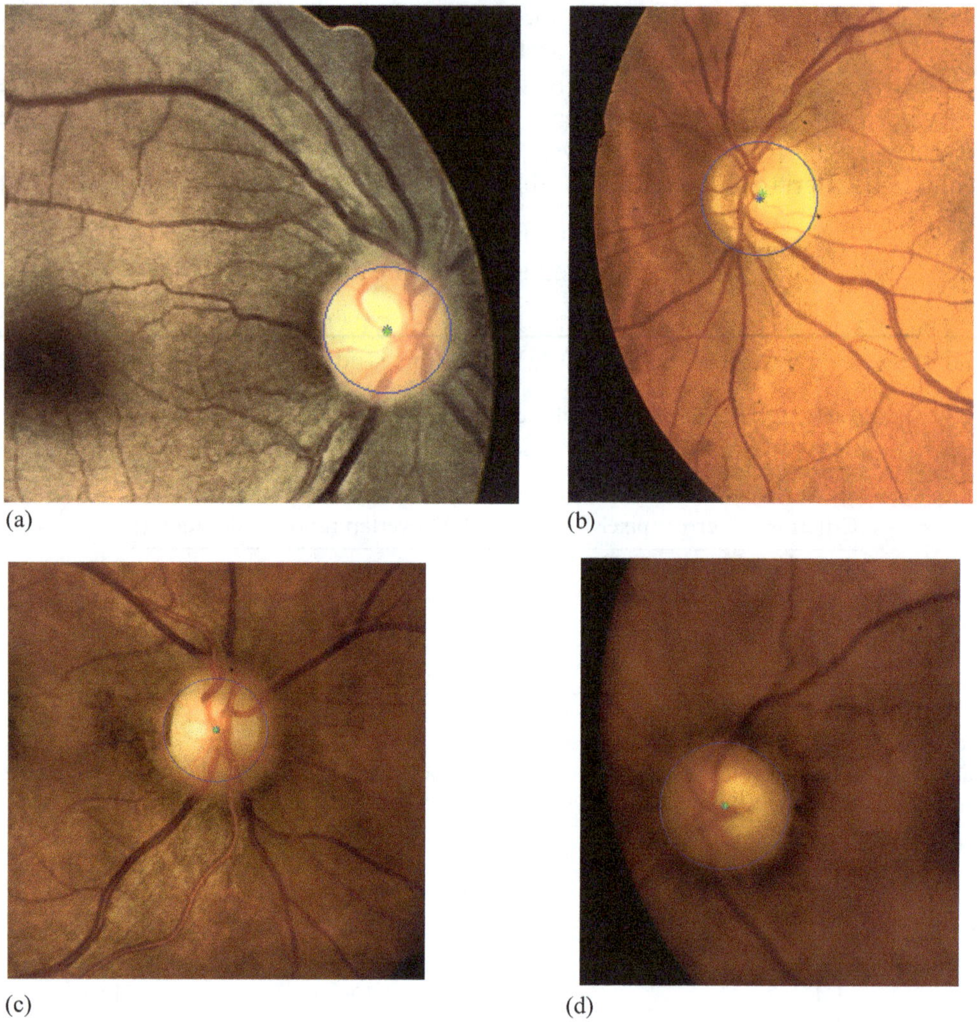

Figure 2.16. Some of the successful OD detection results on a) DRIVE b) STARE c) DIAREDTB0 and d) DIARETDB1. Reproduced with permission from [1]. Copyright 2018 Elsevier.

STARE, DIARETDB0, DIARETDB1, and DRIVE retinal archives are, respectively, set as 55, 78, 104, and 34 pixels of OD center distance error. Some of the well detected OD patterns are shown in figure 2.16 whereas some of the failed results belonging to the STARE and DIARETDB0 retinal archives can be observed in figures 2.17 and 2.18, respectively [1]. The plot color of the OD circle differs for these images because of the changing dominant retinal background colors.

2.4 Discussion and conclusion

The proposed study and the recent state-of-the-art OD detection studies are handled comparatively with respect to the STARE, DIAERTDB0, DIARETDB1, and

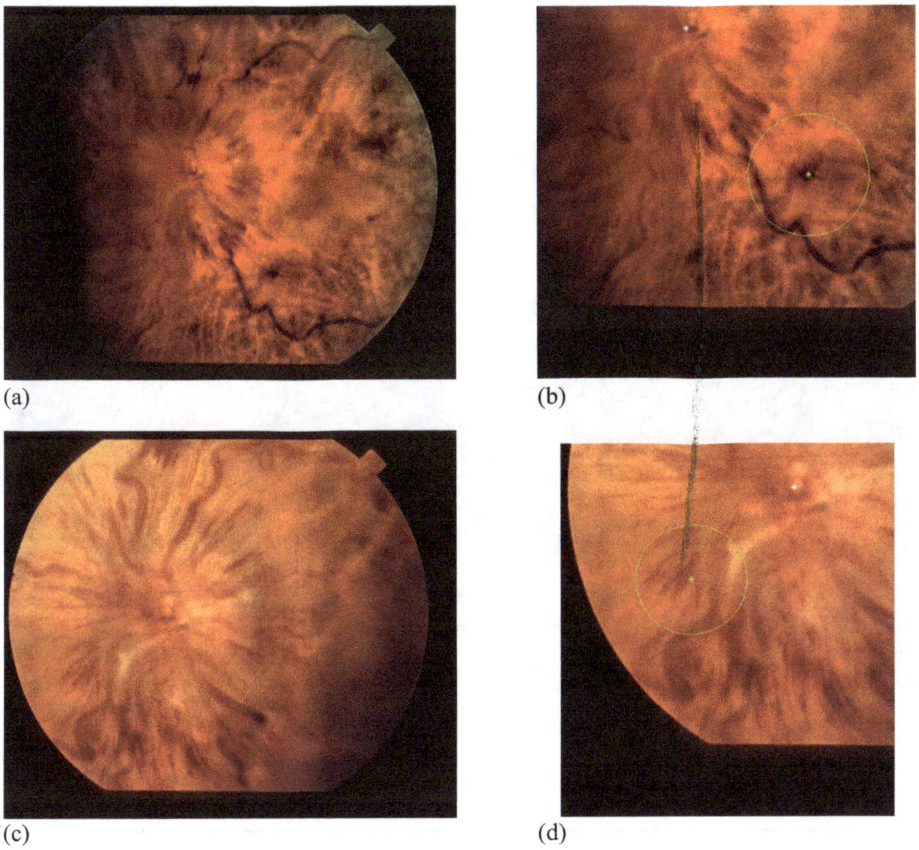

Figure 2.17. Some of the unsuccessful OD detection results on a) 21st image in STARE and b) 21st result in STARE as well as c) 26st image in STARE and d) 26st result in STARE. Reproduced with permission from [1]. Copyright 2018 Elsevier.

DRIVE retinal archives. In this study, the contour coordinates of the DRIVE and center coordinates of DIAERTDB0 and DIARETDB1 of the detected circular patterns are evaluated based on the ground truth annotations provided by two different ophthalmologists. The obtained outcomes related to OD center distance error, execution duration, and OD overlap ratio give promising results.

In the medical image analysis field, in order to obtain more detailed insight about the performance of a proposed method and its reliability compared to doctors, statistics aiming to analyze inter-observer and intra-observer variability has usually been utilized [43–45]. However, inter-observer and intra-observer comparisons cannot be accomplished for this study due to the fact that only one ground truth annotation set is provided for all four retinal archives. Nevertheless, the distributions of the Euclidean OD distance errors are analyzed via the dendogram method, as shown in figures 2.19–2.22 [1]. No matter how many images they have, the group number of the dendogram analysis is set as 20 for all the retinal archives. The horizontal axis represents the OD distance error in pixels whereas the vertical axis

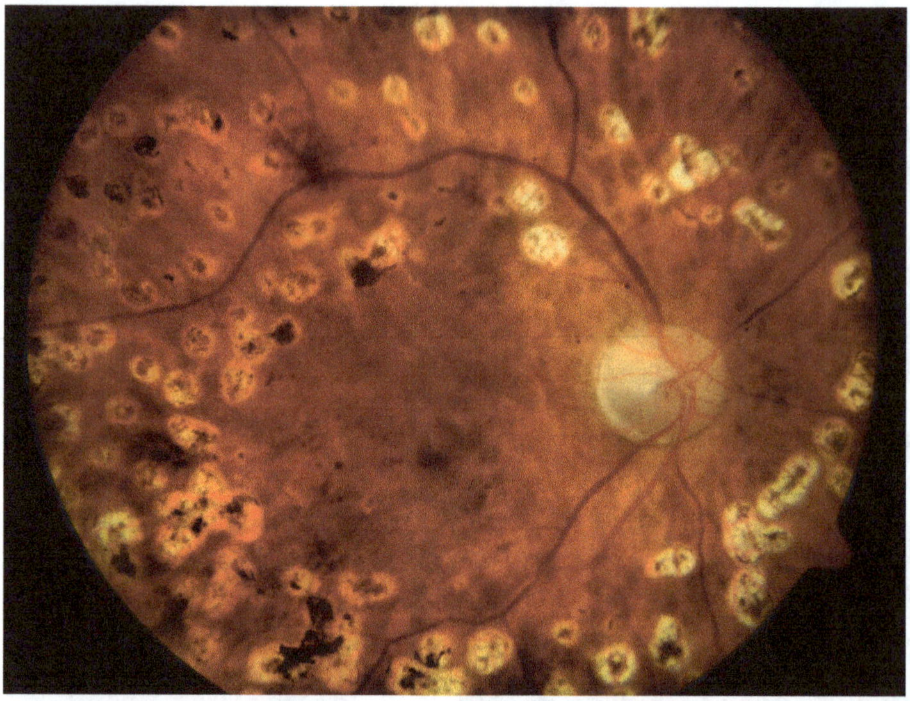

Figure 2.18. Some of the unsuccessful OD detection results on a) 45th image in DIARETDB0 and b) 45th result in DIARETDB0 as well as a) 50th image in DIARETDB0 and b) 50th result in DIARETDB0. Reproduced with permission from [1]. Copyright 2018 Elsevier.

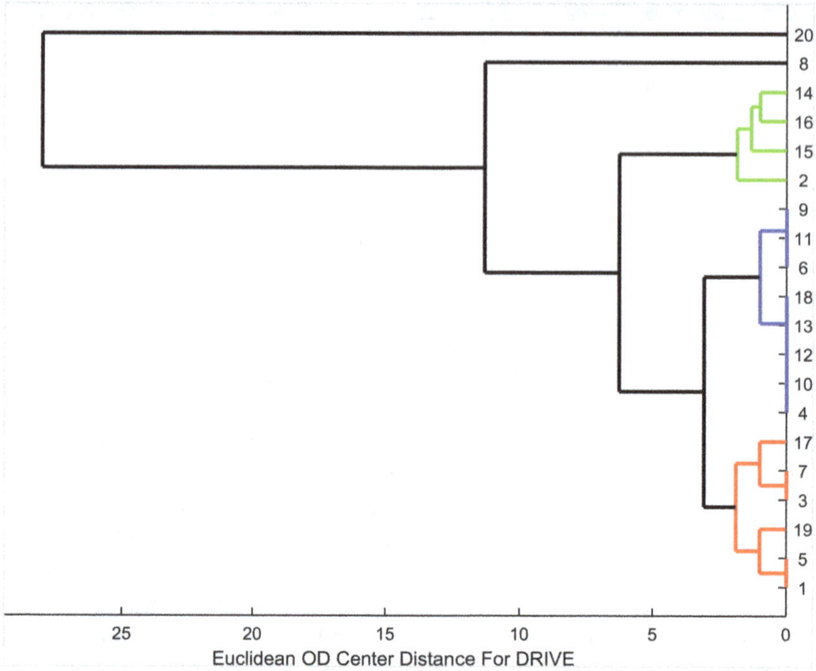

Figure 2.19. The Dendrogram analysis of Euclidean OD center distances of DRIVE is visualized. Reproduced with permission from [1]. Copyright 2018 Elsevier.

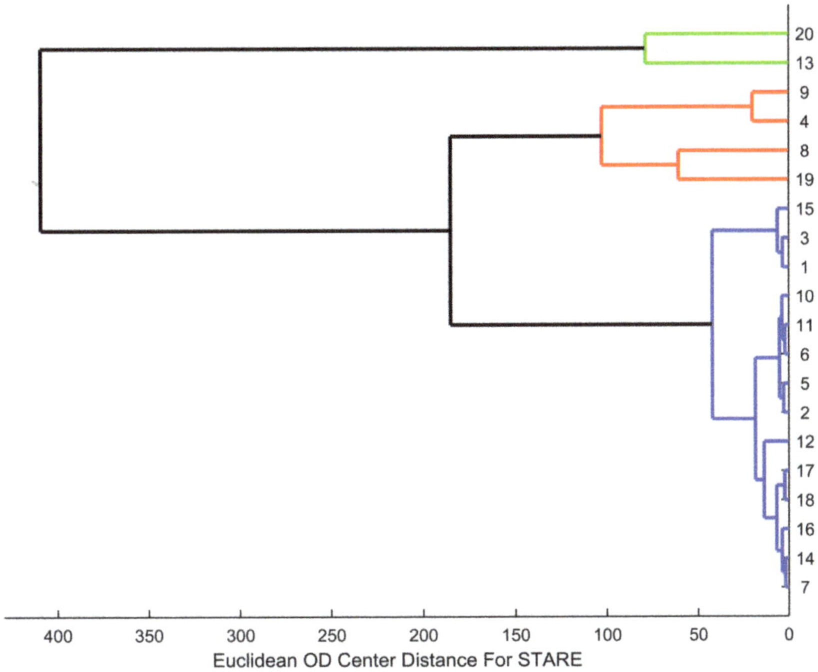

Figure 2.20. The Dendrogram analysis of Euclidean OD center distances of STARE is visualized. Reproduced with permission from [1]. Copyright 2018 Elsevier.

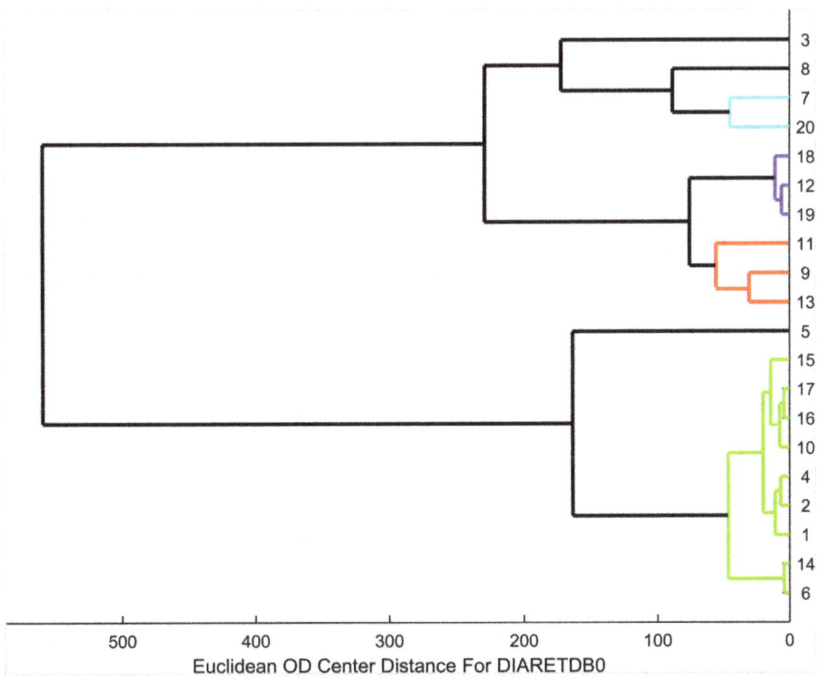

Figure 2.21. The Dendrogram analysis of Euclidean OD center distances of DIARETDB0 is visualized. Reproduced with permission from [1]. Copyright 2018 Elsevier.

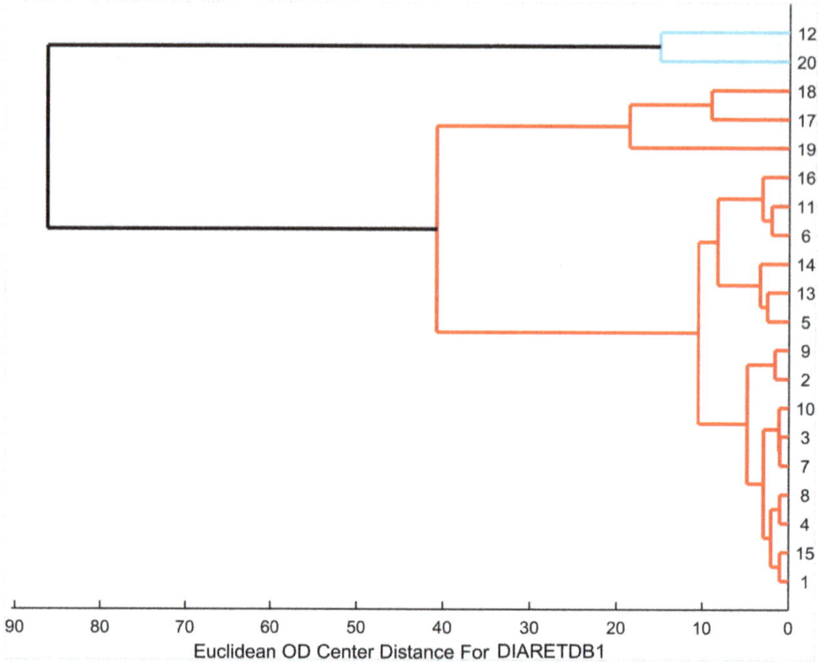

Figure 2.22. The Dendrogram analysis of Euclidean OD center distances of DIARETDB1 is visualized. Reproduced with permission from [1]. Copyright 2018 Elsevier.

represents the number of groups which may have at least one OD center distance error belonging to a unique retinal image. In the DRIVE retinal archive, all the OD center distance errors are in the admissible range, as can be observed in figure 2.19 [1]. There are two outlier groups which have in total three retinal images whose OD distance errors range between 11 and 34 pixels whereas the rest are below 10 pixels and very close to each other. In the STARE retinal archive, the OD center distance errors are in an acceptable range for 74 of 81 retinal images. The group numbers colored in blue represent the OD distance errors which are between 7 and 15 pixels in figure 2.20 [1]. The OD center distance errors are less than 55 pixels and very close to each other. In the DIARETDB0 retinal archive, OD center distance errors are in an admissible range for 117 of 130 retinal images. The group numbers colored in green symbolize the OD distance errors which are between 6 and 15 pixels in figure 2.21 [1]. The OD center distance errors are less than 78 pixels and too similar to each other. In the DIARETDB1 retinal archive, all the OD center distance errors are in an acceptable range, as shown in figure 2.22 [1]. There is one outlier group which has in total three retinal images whose OD distance error ranges are between 61 and 104 pixels, whereas the remaining ones are below 47 pixels and highly similar to each other.

The mean OD center distance error of this study is below the outcomes for most of the previous studies for the DRIVE retinal image archive [2, 28, 29, 46–51]. The mean OD center distance error of this study is below the outcomes for some of the previous studies for the STARE retinal image archive [29, 46, 48, 49]. The mean OD overlap ratio of this study is higher than the results for some previous studies for the DRIVE retinal image archive [28, 47, 48, 51, 52]. The mean OD center distance error of this study is better than the outcomes of most of the previous studies for DIARETDB1 retinal image archive [17, 53–58]. Despite the fact that it is not very reliable to evaluate the execution durations of studies simply by ordering their results, since CPU or GPU capacities highly affect their speed, the mean execution duration of this study is at an admissible position among the previous studies [1].

The former property based studies utilizing only the color, brightness, and circularity of the OD region may be prevented from detecting OD patterns because of the existence of some circule-like and yellowish exudates. However, this study offers a more robust approach by first deciding on an approximate OD location and then confidently applying more detailed steps to find the exact OD center and contours. More and more, the shortcomings of this study are framed as the three main anomaly types for non-standard retinal images and alternative proper solutions are proposed. However, there is still a way to go for the outliers such as in DIARETDB0, which have rare vessel distribution geometries or dense scars after laser surgery.

In this study, the proposed method achieved 91.35%, 100%, 100%, and 90% success rates on the STARE, DRIVE, DIARETDB1, and DIARETDB0 retinal archives [1]. The maximum allowed OD distance errors are set as 55, 34, 104, and 78 pixels for the aforementioned retinal archives, respectively [1]. The mean OD center Euclidean distance errors are 11.82, 6.6, 11.75, and 16.74 pixels while the mean execution durations are 24.58, 13.96, 25.26, and 32.07 s for the STARE, DRIVE, DIARETDB1, and DIARETDB0 retinal archives, respectively [1]. The mean OD overlap ratio for DRIVE is 0.0778 [1]. The proposed method achieved encouraging outcomes by estimating the circular patterns corresponding to OD contours with low execution durations and high accuracies for all the four retinal archives. More precise OD contour detection may be performed using some state-of-the-art segmentation methods as a future work.

Acknowledgments

This chapter has been written reusing the article 'Automated fuzzy optic disc detection algorithm using branching of vessels and color properties in fundus images' which was written by Mehmet Nergiz, Mehmet Akın, Abdulnasır Yıldız, and Ömer Takeş [1]. We extend our thanks to the other authors of the article and Professor Anna L Ells who was an ophthalmologist specializing in the retina at the Alberta Children's Hospital Research Institute, Calgary University, Canada, and Professor Emeritus Rangaraj M Rangayyan who was at the Department of Electrical and Computer Engineering, Calgary University, Canada, for providing the ground truth data of the OD center and contours for the DRIVE retinal image archive.

References

[1] Nergiz M, Akın M, Yıldız A and Takeş Ö 2018 Automated fuzzy optic disc detection algorithm using branching of vessels and color properties in fundus images *Biocybern. Biomed. Eng.* **38** 850–67

[2] Youssif A A H A R, Ghalwash A Z and Ghoneim A A S A R 2008 Optic disc detection from normalized digital fundus images by means of a vessels' direction matched filter *IEEE Trans. Med. Imaging* **27** 11–8

[3] Fundus Photography http://en.wikipedia.org/wiki/Fundus_photography

[4] Allam A M N, Youssif A A-H and Ghalwash A Z 2015 Automatic segmentation of optic disc in eye fundus images: a survey *Electron. Lett. Comput. Vis. Image Anal.* **14** 1–20

[5] WHO Diabetes http://who.int/mediacentre/factsheets/fs312/en/,2017

[6] Muangnak N, Aimmanee P and Makhanov S 2018 Automatic optic disk detection in retinal images using hybrid vessel phase portrait analysis *Med. Biol. Eng. Comput.* **56** 583–98

[7] Rodrigues L C and Marengoni M 2017 Segmentation of optic disc and blood vessels in retinal images using wavelets, mathematical morphology and Hessian-based multi-scale filtering *Biomed. Signal Process. Control* **36** 39–49

[8] Zahoor M N and Fraz M M 2017 Fast optic disc segmentation in retina using polar transform *IEEE Access* **5** 12293–300

[9] Abramoff M D, Garvin M K and Sonka M 2010 Retinal imaging and image analysis *IEEE Rev. Biomed. Eng.* **3** 169–208

[10] Fan Z *et al* 2018 Optic disk detection in fundus image based on structured learning *IEEE J. Biomed. Health Inform.* **22** 224–34

[11] Mookiah M R K, Acharya U R, Chua C K, Lim C M, Ng E and Laude A 2013 Computer-aided diagnosis of diabetic retinopathy: a review *Comput. Biol. Med.* **43** 2136–55

[12] Schaefer G and Clos A 2010 Image analysis for exudate detection in retinal images *Biocomputation and Biomedical Informatics: Case Studies and Applications* ed A A Lazakidou (IGI Global) ch 13 pp 198–203

[13] Abramoff M D, Niemeijer M, Xu X and Sonka M 2012 Automated determination of arteriovenous ratio in images of blood vessels *EP Patent* EP2665406A1

[14] Al-Bander B, Al-Nuaimy W, Williams B M and Zheng Y 2018 Multiscale sequential convolutional neural networks for simultaneous detection of fovea and optic disc *Biomed. Signal Process. Control* **40** 91–101

[15] Tan J H, Acharya U R, Bhandary S V, Chua K C and Sivaprasad S 2017 Segmentation of optic disc, fovea and retinal vasculature using a single convolutional neural network *J. Comput. Sci.* **20** 70–9

[16] Díaz-Pernil D, Fondón I, Peñ-Cantillana F and Gutiérrez- Naranjo M A 2016 Fully automatized parallel segmentation of the optic disc in retinal fundus images *Pattern Recogn. Lett.* **83** 99–107

[17] Alshayeji M, Al-Roomi S A and Abed S 2017 Optic disc detection in retinal fundus images using gravitational law-based edge detection *Med. Biol. Eng. Comput.* **55** 935–48

[18] Dai B, Wu X and Bu W 2017 Optic disc segmentation based on variational model with multiple energies *Pattern Recogn.* **64** 226–35

[19] Hou K, Liu N, Jia W, He Y, Lian J and Zheng Y 2018 Optic disc detection from fundus photography via best-buddies similarity *Appl. Sci.* **8** 709

[20] Lalonde M, Beaulieu M and Gagnon L 2001 Fast and robust optic disc detection using pyramidal decomposition and Hausdorff-based template matching *IEEE Trans. Med. Imaging* **20** 1193–200

[21] Lowell J *et al* 2004 Optic nerve head segmentation *IEEE Trans. Med. Imaging* **23** 256–64

[22] Wang Y, Zhang H and Yang F 2017 A weighted sparse neighbourhood-preserving projections for face recognition *IETE J. Res.* **63** 358–67

[23] Zhang H, Cao L and Gao S 2014 A locality correlation preserving support vector machine *Pattern Recogn.* **47** 3168–78

[24] Grewal P S, Oloumi F, Rubin U and Tennant M T 2018 Deep learning in ophthalmology: a review *Can. J. Ophthalmol* **53** 309–13

[25] Nergiz M and Akın Arı M 2016 A fast optic disc localization algorithm based on geometry of vessel distribution *Int. Conf. on Natural Science and Engineering (ICNASE'16)*

[26] Databases - Images Sciences Institute http://isi.uu.nl/Research/Databases (Accessed: 31 October 2022)

[27] Staal J, Abramoff M D, Niemeijer M, Viergever M A and van Ginneken B 2004 Ridge-based vessel segmentation in color images of the retina *IEEE Trans. Med. Imaging* **23** 501–9

[28] Zhu X, Rangayyan R M and Ells A L 2010 Detection of the optic nerve head in fundus images of the retina using the hough transform for circles *J. Digital Imaging* **23** 332–41

[29] Rangayyan R M, Zhu X, Ayres F J and Ells A L 2010 Detection of the optic nerve head in fundus images of the retina with Gabor filters and phase portrait analysis *J. Digital Imaging* **23** 438–53

[30] STARE: Locating the optic nerve http://cecas.clemson.edu/~ahoover/stare/nerve/index.html (Accessed: 2 November 2022)

[31] Hoover A D, Kouznetsova V and Goldbaum M 2000 Locating blood vessels in retinal images by piecewise threshold probing of a matched filter response *IEEE Trans. Med. Imaging* **19** 203–10

[32] DIARETDB1 - Standard Diabetic Retinopathy Database http://it.lut.fi/project/imageret/diaretdb1/ (Accessed: 2 November 2022)

[33] Kauppi T, Kalesnykiene V, Kámáráinen J, Lensu L, Sorri I, Raininen A, Voutilainen R, Hannu U, Kámáráinen H and Pietilá J 2007 Diabetic retinopathy database and evaluation protocol *British Machine Vision Conf.*

[34] DIARETDB0 - Standard Diabetic Retinopathy Database http://it.lut.fi/project/imageret/diaretdb0/ (Accessed: 2 November 2018)

[35] Kauppi T, Kalesnykiene V, Kamarainen J-K, Lensu L, Sorri I, Uusitalo H, Kälviäinen H and Pietilä J 2006 DIARETDB0: evaluation database and methodology for diabetic retinopathy algorithms *Technical Report*

[36] Nergiz M, Akın M and Arı Ş 2013 Diyabetik retinopati tespitinde yeni bir algoritma kullanılarak optik disk yerinin kestirimi *Dicle Üniv. Mühendislik Dergisi* **2** 109–20

[37] Nergiz M, Akın M and Arı Ş 2014 Optic disc detection in retinal images via a new algorithm *TıpTekno* 74–7

[38] Zuiderveld K 1994 Contrast limited adaptive histogram equalization *Graphics Gems* ed P S Heckbert vol 4 (San Diego, CA: Academic) pp 474–85

[39] Frangi A F, Niessen W J, Vincken K L and Viergever M A 1998 Multiscale vessel enhancement filtering *Medical Image Computing and Computer-Assisted Intervention* ed W M Wells, A Colchester and S Delp (Berlin: Springer) pp 130–7

[40] Nergiz M and Akın M 2017 Retinal vessel segmentation via structure tensor coloring and anisotropy enhancement *Symmetry* **9** 276–84

[41] Reza A M 2004 Realization of the contrast limited adaptive histogram equalization (CLAHE) for real-time image enhancement *J. VLSI Signal Process. Syst. Signal Image Video Technol.* **38** 35–44

[42] Otsu N 1979 A threshold selection method from gray-level histograms *IEEE Trans. Syst. Man Cybern.* **9** 62–6

[43] Santos B S, Ferreira C, Silva J S and Teixeira A S L 2004 Quantitative evaluation of a pulmonary contour segmentation algorithm in x-ray computed tomography images *Acad. Radiol.* **11** 868–78

[44] Antunes S G, Silva J S, Santos J B, Martins P and Castela E 2011 Phase symmetry approach applied to children heart chambers segmentation: a comparative study *IEEE Trans. Biomed. Eng.* **58** 2264–71

[45] Silva J S, Santos J B, Roxo D, Martins P, Castela E and Martins R 2012 Algorithm versus physicians variability evaluation in the cardiac chambers extraction *IEEE Trans. Inf. Technol. Biomed.* **16** 835–41

[46] Dehghani A, Moin M-S and Saghafi M 2012 Localization of the optic disc center in retinal images based on the Harris corner detector *Biomed. Eng. Lett.* **2** 198–206

[47] Harangi B and Hajdu A 2015 Detection of the optic disc in fundus images by combining probability models *Comput. Biol. Med.* **65** 10–24

[48] Mahfouz A E and Fahmy A S 2010 Fast localization of the optic disc using projection of image features *IEEE Trans. Image Process.* **19** 3285–9

[49] Mithun N C, Das S and Fattah S A 2014 Automated detection of optic disc and blood vessel in retinal image using morphological, edge detection and feature extraction technique *16th Int. Conf. on Computer and Information Technology (ICCIT)* 98–102

[50] Qureshi R J, Kovacs L, Harangi B, Nagy B, Peto T and Hajdu A 2012 Combining algorithms for automatic detection of optic disc and macula in fundus images *Comput. Vis. Image Understand.* **116** 138–45

[51] Welfer D, Scharcanski J and Marinho D 2013 A morphologic twostage approach for automated optic disk detection in color eye fundus images *Pattern Recogn Lett.* **34** 476–85

[52] Niemeijer M, Staal J, van Ginneken B, Loog M and Abramoff M D 2004 Comparative study of retinal vessel segmentation methods on a new publicly available database *Proc. SPIE* **5370** 5370–9

[53] Bharkad S D 2016 Automatic segmentation of optic disc in retinal images using dwt *IEEE 6th Int. Conf. on Advanced Computing (IACC)* 386–91

[54] Nimbarte N and Mushrif M 2016 Extraction of blood vessels and optic disc in retinal images *Comput. Methods Biomech. Biomed. Eng.: Imaging Visual.* **6** 31–42

[55] Rahebi J and Hardalaç F 2016 A new approach to optic disc detection in human retinal images using the firefly algorithm *Med. Biol. Eng. Comput.* **54** 453–61

[56] Xiong L and Li H 2016 An approach to locate optic disc in retinal images with pathological changes *Comput. Med. Imaging Graph.* **47** 40–50

[57] Akyol K, Şen B and Bayir A 2016 Automatic detection of optic disc in retinal image by using keypoint detection, texture analysis, and visual dictionary techniques, computational and mathematical methods in medicine **2016** 6814791

[58] Basit A and Fraz M M 2015 Optic disc detection and boundary extraction in retinal images *Appl. Opt.* **54** 3440–7

IOP Publishing

Photo Acoustic and Optical Coherence Tomography Imaging, Volume 2
Fundus imaging for the retina
Ayman El-Baz and Jasjit S Suri

Chapter 3

Application of enface image registration/alignment to introduce new ocular imaging biomarkers

Zahra Amini, Rahele Kafieh, Mahnoosh Tajmirriahi, Zeinab Parsons and Hossein Rabbani

This chapter is dedicated to discussing the application of enface image registration/alignment to introduce new ocular imaging biomarkers. The enface images discussed in this chapter are colour fundus image, scanning laser ophthalmoscopy (SLO), fundus fluorescein angiography (FFA), and enface optical coherence tomography (OCT) thickness maps. In addition to the different features that can be extracted from these modalities, new ocular imaging biomarkers are also introduced based on the registration/alignment of these enface images.

3.1 Introduction

Fundus imaging is considered the most well-known ocular enface imaging method, with the extraction of biomarkers which facilitates the diagnosis and treatment of related diseases. The other enface modalities, on the other hand, provide complementary information about different ocular structures, which are non-negligible in the comprehensive diagnosis of related retinal diseases. A main step before the detection of informative biomarkers, however, is the registration and alignment of such a group of images. Accordingly, effective anatomic landmarks of the eye and different enface imaging modalities are first introduced. Then we discuss the applications of enface image registration and alignment to introduce new ocular imaging biomarkers. For this purpose, feature extraction techniques from retinal enface images are elaborated in section 3.2. The process of registration and alignment of optical coherence tomography (OCT) and fundus images in the right

doi:10.1088/978-0-7503-2056-6ch3

and left eyes based on the features mentioned is then discussed in section 3.3. Section 3.4 is devoted to methods of registering between fundus fluorescein angiography and scanning laser ophthalmoscopy images. Accordingly, new ocular imaging biomarkers are discussed in the conclusion.

3.1.1 Anatomic landmarks of the eye

In this subsection, we mostly concentrate on the most effective ocular landmarks applicable in the registration and alignment of ocular images. These landmarks are demonstrated in figure 3.1. The optic disc (optic nerve head (ONH)) as one of these landmarks is the point in the retina where the axons of ganglion cells leave the eye forming the optic nerve (figures 3.1(a)–(c)). Since there are no rod or cone cells at this point, a break in the visual field is created here, called the blind spot. The position of the blind spot is about 15° nasal to the fovea. There is a dark region in the macula empty of retinal blood vessels, which is known as the foveal avascular zone (FAZ) (figure 3.1(e)). There is a small rod-free depression in the center of the FAZ within the macula which is associated with color vision and perception of fine details due to the high concentration of cone cells. This small pit is called the fovea. The absence of blood vessels in the FAZ is thought to be responsible for reducing light scattering in the fovea and maximizing the optical quality. The fovea and FAZ are two of the most effective landmarks as well. More biological details are given in [1].

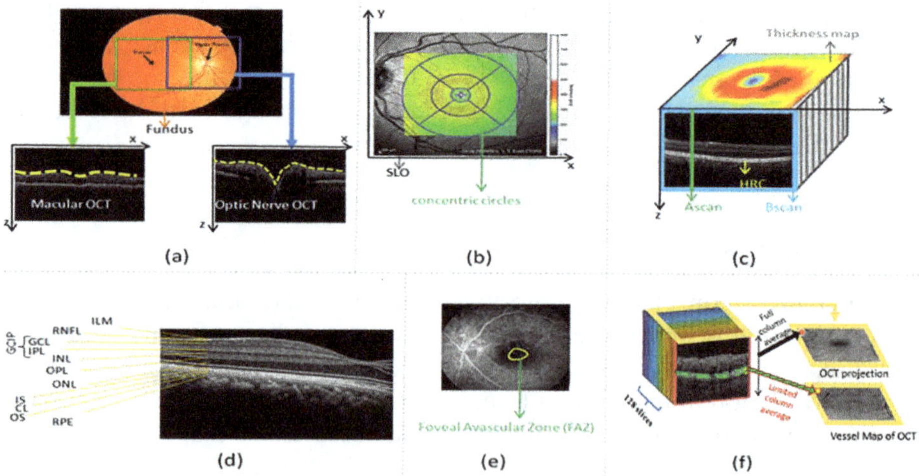

Figure 3.1. Anatomic landmarks of the eye. (a) Fundus image with main landmarks: fovea and optic nerve head (ONH), and obtained OCTs from the macular and ONH regions. (b) SLO image. (c) 3D OCT and related parts. (d) Retinal layers and boundaries on the macular OCT. (e) Foveal avascular zone (FAZ) on the FFA. (f) Construction of projection and vessel map from 3D OCTs. ((a)–(c) Reproduced with permission from [2]. Copyright 2019 Elsevier.)

3.1.2 Enface image modalities

Many diseases such as eye and systemic diseases can be identified and followed from retinal changes. Since the process of vision is associated with several anatomical structures, this subsection is devoted to retinal imaging for expressing these structures. Various retinal imaging modalities are demonstrated in figure 3.2. A piece of multifaceted software is already developed to analyze most of the above mentioned images comprehensively [2]. However, the focus of this work is on enface retinal images (fundus imaging, scanning laser ophthalmoscopy (SLO), and fundus fluorescein angiography (FFA)), elaborated below.

3.1.2.1 Fundus imaging

In fundus imaging process, reflected light from the semi-transparent tissues of the retina is projected onto the imaging plane and a 2D projection of those 3D tissues is obtained. Thus, fundus imaging comprises any process resulting in a 2D image with intensities corresponding to the amount of reflected light. According to this definition, OCT imaging cannot be categorized as a fundus imaging technique. Fundus imaging includes fundus photography (including so-called red-free photography), color fundus photography, stereo fundus photography, and hyper spectral imaging [5].

3.1.2.2 Scanning laser ophthalmoscopy

SOL method is a non-invasive modality for imaging the retina using a different approach from the fundus camera [6]. In this method a laser beam from a near-infrared diode is swept over the retina. A confocal photodiode that is conjugate to the retinal plane detects the reflected light and the retinal image is obtained.

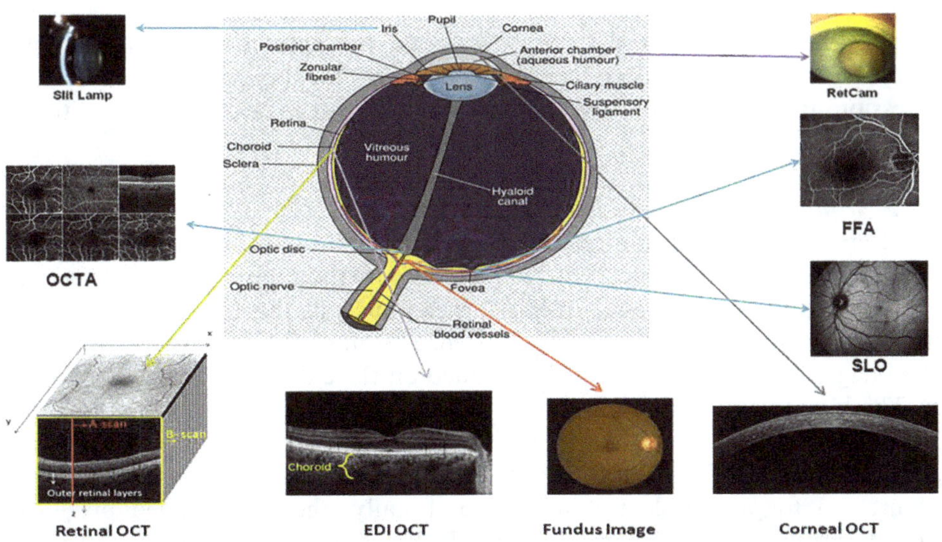

Figure 3.2. Ocular imaging modalities. (Reproduced with permission from [4]. Copyright 2016 Wiley.)

Since the SLO can image structures at a very high magnification and high frame rate, it produces a higher contrast retinal image than fundus imaging while using lower levels of light exposure. In addition, a topographic 3D map of the retina can be made from images obtained from the SLO to estimate retinal thickness. Such specifications allow accurate diagnosis of retinal structures and objective evaluation of the optic disc in some disorders such as glaucoma.

3.1.2.3 Fundus fluorescein angiography

FFA relies on illumination [7, 8]. In this modality a special fluorescein dye is injected intravenously into the veins within the antecubital fossa and is carried around by the blood stream. Fundus images are taken at intervals, starting at the time of injection. During imaging, a blue absorption filter is applied to the white light emitted from the fundus camera and the output of this filter enters the eye. This light is absorbed by the unbound fluorescein flowing in the vessels of the eye. Then emitted photons from fluorescing molecules are passed through a barrier filter and the yellow–green part of the light spectrum is extracted and detected to produce a corresponding image. This technique is widely used to precisely figure out the perfusion status of the retina at the posterior pole and in the far periphery and to diagnose retinal vascular diseases.

3.2 Feature extraction

In order to register or to align ocular images, it is necessary to detect guiding markers through feature extraction. Such features may be categorized into the main anatomic landmarks such as the retinal vessels, ONH, and fovea. For this purpose, we present dominant sample methods to enable automatic detection of such features in the following subsections.

3.2.1 Vessel segmentation

Segmentation of vessels of the retina leads to the extraction of morphological attributes which can be utilized as effective features for the diagnosis, checking the progression, treatment, and assessment of multiple cardiovascular and ophthalmologic diseases.

3.2.1.1 Segmentation of vessels in fundus images

An efficient method for the automatic segmentation of blood vessels in color fundus images was proposed in [9] and used in [10]. This method improved the detection of low contrast and narrow vessels using curvelet analysis through four steps. In the first step the contrast of the image was enhanced using modified curvelet coefficients of the image, then match filtering was applied on the enhanced image in the second step. The blood vessels were intensified through match filtering. In the third step a curvelet transform was used to distinguish vessels from its background. In this way edges were extracted. The results of different stages of this method are demonstrated in figure 3.3 for a sample fundus image. Finally, the misclassified pixels were removed through length filtering. The DRIVE dataset [11] was used to assess the performance of this method compared with a hand-labeled ground truth.

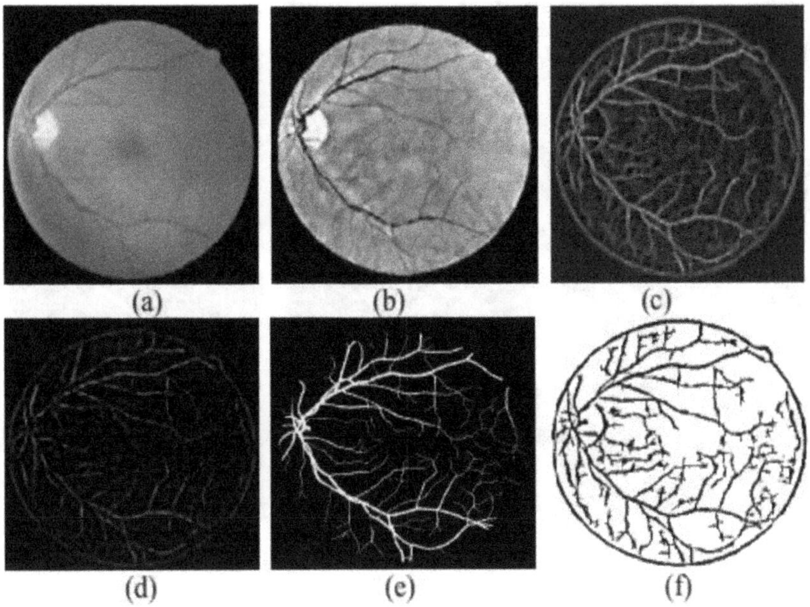

Figure 3.3. Segmentation of vessels in fundus images. (a) Input image, (b) enhanced image, (c) match filtered response on (b), (d) low frequency components, (e) manual segmentation, and (f) result of [9]. (Reproduced with permission from [9]. Copyright 2009 IEEE.)

Since curve discontinuities can be handled well by a curvelet transform, this method achieved an accuracy of 0.9631 which demonstrates the superiority of this method in comparison to known techniques.

3.2.1.2 Segmentation of vessels in FFA images

A novel algorithm was presented in [12] to extract the blood vessels centerline in FFA images and to determine different stages of diabetic retinopathy. There were six steps in this algorithm: variations of background intensity were reduced by using a nonlinear diffusion method in the first step. In the second step a morphological edge detector was utilized to enhance the result images of the first step. In the third step the curvelet transform was applied to produce a set of directional images with enhanced edges. In the fourth step, a Hessian matrix and first order derivative of the directional images were used to individually process each of these images in a multi-scale framework and to extract initial centerline segments. In the fifth step length, intensity features and eigenvalue analysis were performed on every resulting segment in the previous step to confirm or reject it. Finally, image subsets were connected in a binary image and the centerline segmentation of the vessels was obtained. The result of this algorithm is demonstrated and compared with manual segmentation in figure 3.4. The performance of this algorithm was evaluated by applying it through 70 FFA images from different stages of diabetic retinopathy and a true positive ratio of 0.9017 and a false positive ratio of 0.983 were obtained.

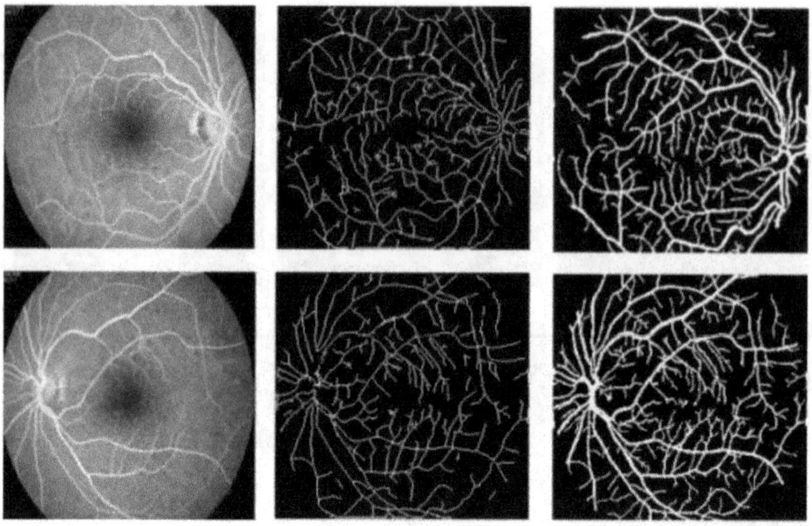

Figure 3.4. (a) FFA images, (b) vessel centre line, and (c) manual segmentation. (Reproduced with permission from [12]. Copyright 2013 IEEE.)

3.2.2 Optic disc segmentation

In an enface image, the optic disc (OD) and exudates vary in their amount of brightness, contrast, and color between different patients. This fact can be used as a landmark to determine the severity of retinal diseases. For this purpose an automatic algorithm was proposed in [13] in order to detect the OD in images and particularly in low contrast ones. This method was based on a digital curvelet transform (DCUT). The superiority of this method was that user initialization was not necessary and the method was robust to the appearance changes of retinal fundus images. This method consisted of three main stages. First, a contrast limited adaptive histogram equalization algorithm was used in the red and green planes to enhance the contrast of the OD. In the second step DCUT was applied to the enhanced image then modified curvelet coefficients were employed to extract bright lesions which are candidates for the OD in the image. Then, a new OD detection method was used based on the DCUT and boundary extraction method level set method. In the final stage, the bright lesions map (BLM) was generated then exudates and the OD were distinguished from each other. Finally, the OD region was determined and other extracted pixels of the BLM which were out of the OD region were categorized as bright lesions. Figure 3.5 illustrates sample results of this method. This method was applied to the 89 images of DIARETDB1 and achieved an average accuracy of 94.51% in the extraction boundaries of the OD. Later, a fast method for the detection of the OD was proposed in [14]. This method was based on correlation and least squares and obtained an accuracy of 98% in the detection of OD.

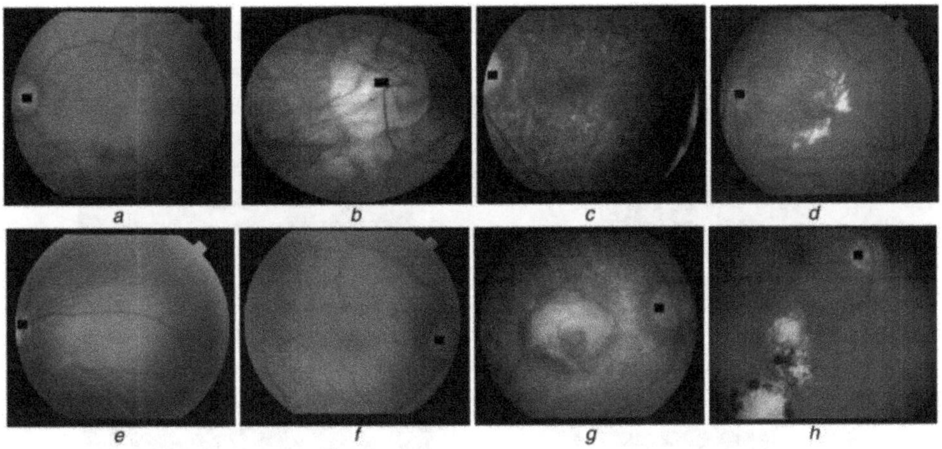

Figure 3.5. Sample results of OD detection based on the method in [13]. (Reproduced with permission from [13]. Copyright 2012 Wiley.)

3.2.3 FAZ segmentation

Diabetic retinopathy can cause blindness in its advanced stage. Thus regular screening is necessary to achieve early detection of diabetic retinopathy using retinal images. There is no uniform distribution for DR abnormalities over the retina and certain types usually occur in particular regions of the retina. A useful feature for grading diabetic retinopathy is the distance between lesions and the FAZ. The authors of [3] found the location of the FAZ in FFA images utilizing a new fully automatic system. Their proposed method was based on two main stages: a DCUT and morphological operations. In the first step a vessel segmentation based on DCUT was applied to detect the end points of the vessels. These end points were connected in the selected region of interest to extract the FAZ region. In the second stage morphological operations such as dilatation and erosion were applied on the image resulting from the subtraction of vessels from the retinal image. Using a suitable threshold, the FAZ region were detected in this stage. In the final stage a logical AND was performed between two segmented FAZs then the final FAZ region was extracted. Sample results of FAZ detection using this method are demonstrated in figure 3.6. This method was evaluated for normal stage, mild/moderate non-proliferative diabetic retinopathy, and severe non-proliferative diabetic retinopathy; achieving specificity and sensitivity of >98% and >96%, >98% and >95%, and >97% and >93%, respectively.

3.3 Registration of fundus and OCT images

In recent years, automatic diagnosis systems for various diseases have been significantly improved based on advanced analysis of retinal images. Using these devices, in addition to saving time, a lot of money is also saved. Various changes of retinal pathologies and abnormalities of retinal thickness can be detected by the new techniques of 3D OCT imaging. In addition, for several years retinal

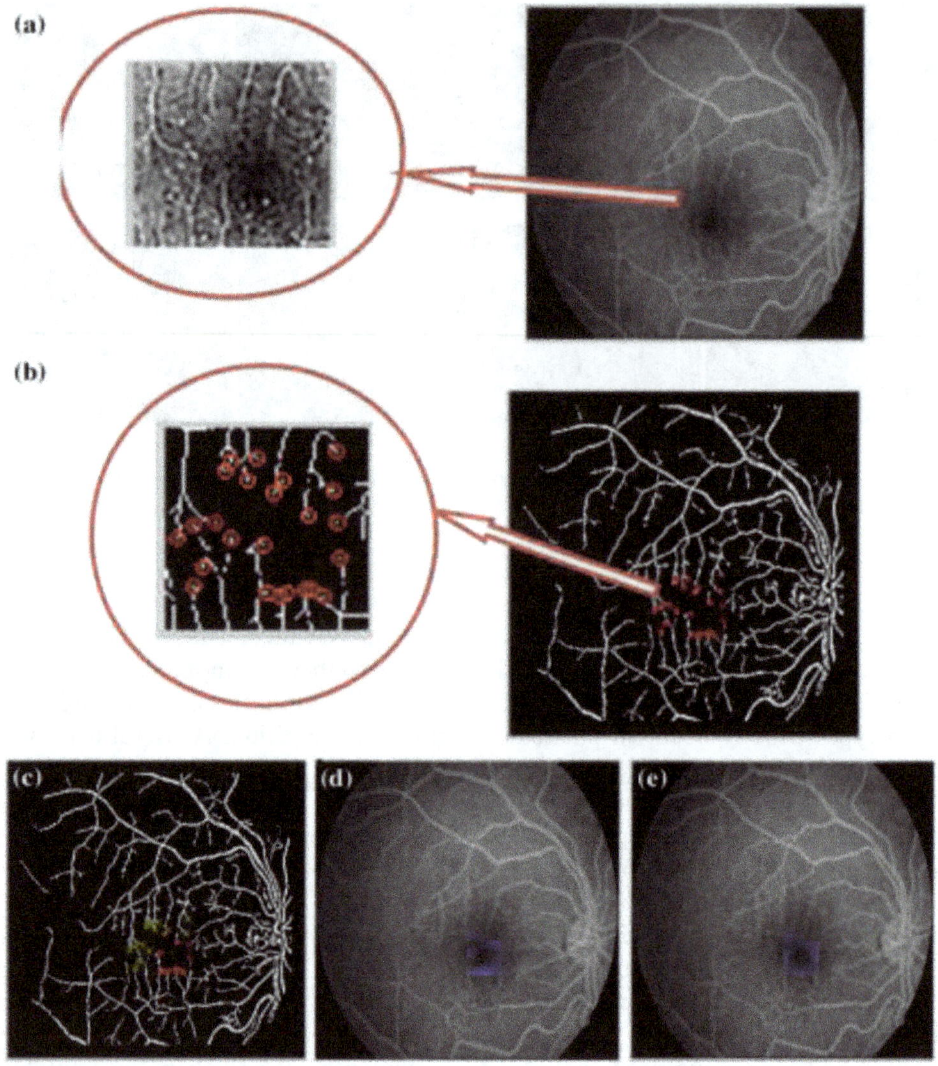

Figure 3.6. Sample result of FAZ detection. (a) Original image, (b) extracted red points, (c) removing unnecessary points, (d) selected red points, and (e) FAZ segmentation based on [3]. (Reproduced with permission from [3]. Copyright 2014 Springer.)

abnormalities have been detected via fundus color images. Thus, through the combination of two imaging modalities more informative images will be obtained. For example, color fundus images represent some abnormalities such as drusen, geographic atrophy, and macular hemorrhages, while OCT images provide their morphology and position. However, registration of color fundus images with OCT projection is the first step to combine these modalities. Automatic registration of 3D OCT data with enface images is already implemented by some vendors, such as Heidelberg Engineering which uses SLO as enface data. However, this remains a

Figure 3.7. Sample registration of vessels on a fundus image based on [10]. (Reproduced with permission from [10]. Copyright 2012 IEEE.)

challenging point when we consider fundus data. In the study demonstrated in [10], a curvelet transform was used to extract vessels from two modalities of 3D OCT and fundus image, then the extracted vessels were registered together using a search algorithm. In the search algorithm images were expanded and contracted in the X and Y directions and the similarity measures between them were computed keeping the center of images at a fixed point. In this study the Topcon 3D OCT device was used to acquire images from ten eyes. This device can observe the retina, and take fundus images and tomograms. Furthermore, it can also provide a corresponding color fundus image. A grayscale fundus image was extracted from the OCT data and the vessels of both the color and grayscale fundus were registered together. Figure 3.7 shows sample registration of vessels on a fundus image based on this method. The performance of this method was evaluated by calculating the mean and standard deviation value of the maximum similarity measure. The results showed a maximum similarity measure of 0.92 ± 0.1 across ten subjects.

3.4 Registration and alignment of ocular images in the right and left eyes

3.4.1 Symmetry analysis of vessel maps in the right and left eyes

Symmetry in anatomical features is observed in most healthy paired tissues. Normal tolerance limits for asymmetry of retinal features in the left and right eye can be used as a marker to detect abnormalities. For example, the asymmetric pattern of the vascular map in scans of both eyes determines diabetic retinopathy, so analysis of asymmetries can provide an opportunity for early detection of diabetic retinopathy but a quantitative comparison is a challenging point and needs effective biomarkers. In a recent study presented in [15] the vascular patterns of both eyes were compared by utilizing a combination of fractal analysis and random transformation [16, 17].

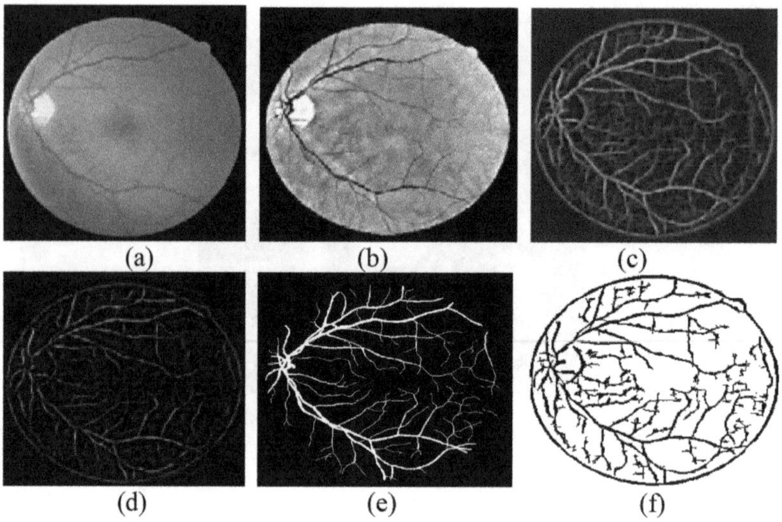

Figure 3.8. Sample results from [16]. (a) The original fundus image, (b) segmented vessels, (c) skeletonization, and (d) thickness map. (Reproduced with permission from [16]. Copyright 2015 IEEE.)

This analysis provided the statistical and geometrical distribution of vessel thickness. A method proposed in [16] consisted of five steps. In the first step, sub-bands of a fast DCT (FDCT) were applied to the fundus image and vessels were segmented. In the second step, thinning of the binary vessel segmentation was done and the center lines of the vessels were determined. In the third step the thickness was computed for each vessel point and the vessel thickness map (VTM) was acquired. Then, in the next step a radon transform (RT) was applied and the projection of vessel thickness across 0° to 180° was obtained. For multiple versions of fundus images of both eyes, such as segmented vessels and skeletonized vessels, the fractal dimension (FD) and radon transform were computed. The mean and standard deviation values of the FD differences and their corresponding p-values were calculated for fifty normal fundus images in both eyes. The results of this method confirmed its superiority in the differentiation of two eyes from each other. Thus, VTM can be considered as a powerful feature to analyze the amount of symmetry of fundus images. Sample results of different stage of this method are depicted in figure 3.8. The evaluation results of the method showed that for a normal subject, there exists a difference of 0.33 ± 0.11 between the FD of the VTM's RT in both eyes.

3.4.2 Symmetry analysis of macular OCTs in the right and left eyes

Thickness measurements of eleven intra-retinal layers and total retina obtained from retinal OCT images can be used to determine the limits of the normal interocular asymmetry in retinal layers around the macula. For this purpose thickness maps (TMs) of the left and right eyes were calculated in [18]. In this study, in the first step the intra-retinal layer in 3D OCT data was segmented utilizing the method proposed

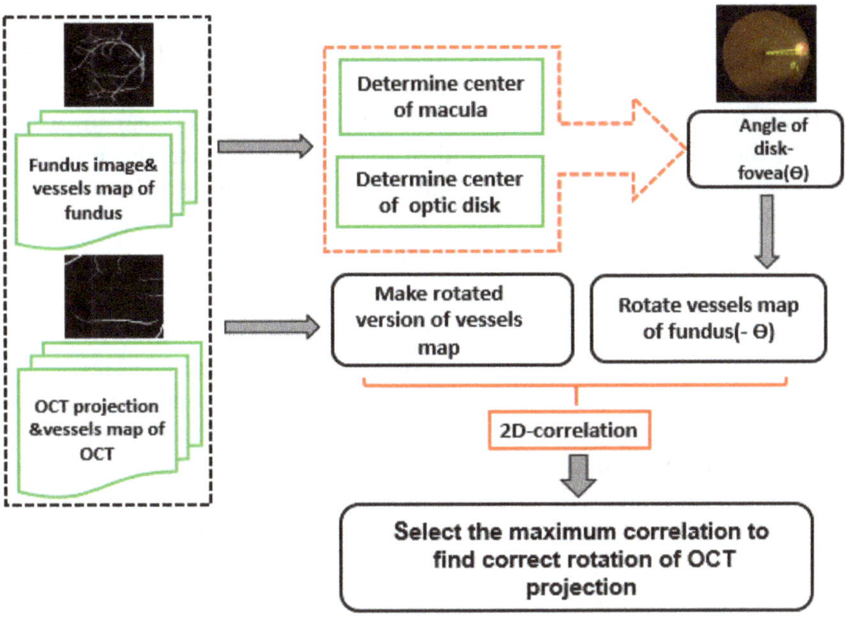

Figure 3.9. Block diagram of the alignment method in the right and left eyes.

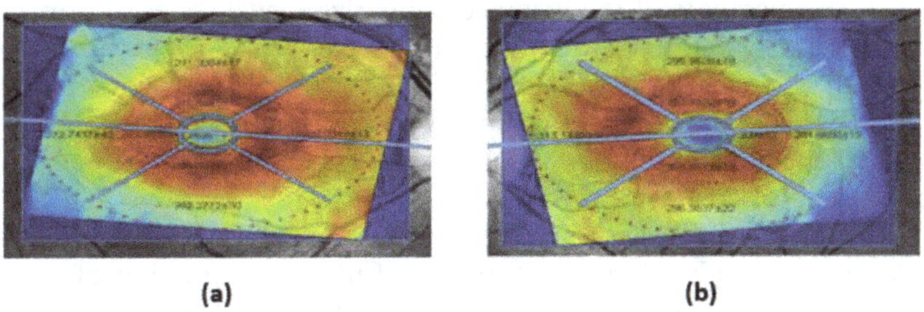

Figure 3.10. Sample registration of OCT and the fundus of (a) the right eye and (b) the left eye based on the method in [18].

in [19] and the mid-macular point was calculated. Then, the TM of each layer was obtained in nine sectors in concentric circles around the macula by the same method as proposed in [20, 21]. In the next step, retinal raphe, which is the axis of the disc–fovea, was aligned and the TMs of the left and right images were registered. This provided a way of comparing corresponding sectors in both eyes. However, the retinal raphe of macular OCTs is not calculable due to limited region size, so the retinal raphe of the fundus images were first aligned and then registration of the OCTs was done to the aligned fundus images. A block diagram of the alignment method is provided in figure 3.9. This method was applied to fundus images of 50 normal subjects. A sample registration result is shown in figure 3.10. The mean and standard deviation of thickness in nine sectors of 11 layers were calculated to analyze the asymmetry in each retinal layer. Significant symmetry was observed in some sectors of retinal layers with $p < 0.05$ in the normal population. By means of this information, the tolerance limits for normal subjects were calculated. Overall, the results of their study demonstrated that there was no identical information in both eyes in the normal population and normal asymmetry in the information of both eyes can be considered as retinal features to diagnose disorders.

3.4.3 Alignment of ONH OCTs in the right and left eyes

New biomarkers can be introduced from OCT images by comparing the symmetry of both eyes in enface images. These biomarkers can be used in the early diagnosis of eye disorders [18, 22]. In the recent study presented in [23] the local cup-to-disc ratio (CDR) was estimated by utilizing aligned OCT B-scans and fundus images and used in analyzing the symmetry of two eyes. This method consisted of two main phases: OCT, fundus image alignment (figure 3.11), and CDR calculation (figure 3.12). In the first phase of this method, the fovea–ONH axes of the left and right fundus images were detected and used to align the corresponding fundus images. Then, the similarity between the OCT B-scan and fundus image was considered to register them. Since the alignment procedure was highly dependent on correct determination of the macula and estimation of the ONH center, two separate methods were used for exact localization of the macula. For this purpose, the end points of the vessels in the FAZ were extracted by utilizing a curvelet-based method and used to determine the center of the macula and the distance regularized level set evolution (DRLSE) algorithm was utilized to find the disc contour of ONH. The second phase of the method was to calculate the local CDR. Since the disc is surrounded by an inner limiting membrane (ILM) and retinal pigment epithelium (RPE) layers, its boundary was determined from these layers through a ridgelet transform in each B-scan. The cup depth was considered the distance between the disc edge and the deepest point of the ILM layer and calculated by extracting near to the horizontal direction of the ridgelet sub-bands of the OCT B-scan. Finally, summation of the areas of cups and summation of the areas of discs are calculated in corresponding local regions and their ratio was considered as the local CDR. In this method, a new quantitative index, namely local volumetric CDR (VCDR), was introduced to assess the symmetry of the 3D OCTs of both eyes in 24 volumetric sectors. This index was calculated by the division of the volume of the cup to its relevant volume of disc. This method was applied to a dataset of forty healthy OCT and their corresponding fundus images to find the maximum symmetry in the upper, middle, and lower regions. The results showed that maximum similarity is in the middle region.

3.5 Registration of FFA and SLO images

3.5.1 FFA and OCT registration using SLO images

As mentioned in section 3.3, the Heidelberg apparatus automatically registers 3D OCT data with enface images (SLO). In the study in [24] this feature is utilized as an intermediate step for automatically registering FFA images with OCT B-scans. They presented a non-rigid correlation-based registration algorithm containing two

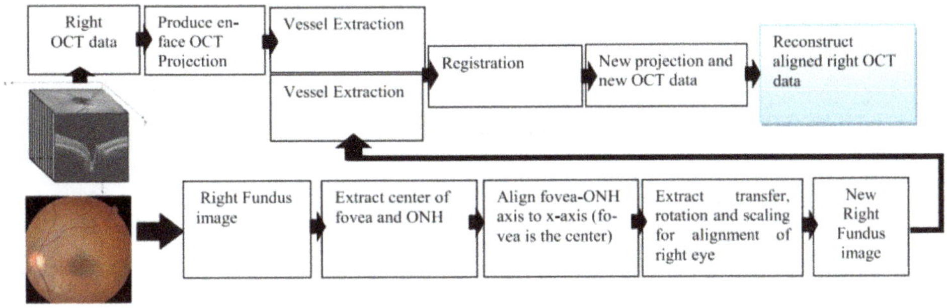

Figure 3.11. Algorithm in [23] for the alignment of OCTs in the right and left eyes. (Reproduced with permission from [23]. Copyright 2019 Elsevier.)

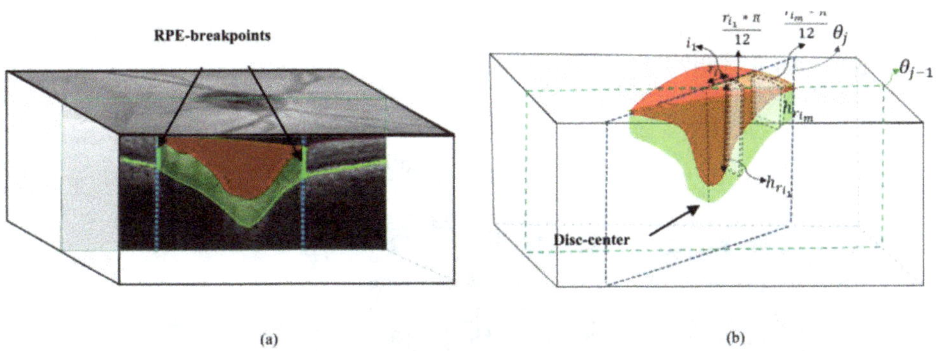

Figure 3.12. Calculation of VCDR in [23]. (a) A sample radial scan and (b) the volume of the cup and disc. (Reproduced with permission from [23]. Copyright 2019 Elsevier.)

registration phases, named the rigid/global and non-rigid/local registrations. In global registration the global parameters for the FFA transformation are optimized and in non-rigid local registration the local parameters are optimized. The block diagram of the method is shown in figure 3.13.

For global registration the vessels are extracted from FFA and SLO images and the results are aligned. The result of global registration is not as perfect as expected, therefore local registration is applied afterwards. The local registration is performed in a batch-wise manner on a region of interest (ROI) of the retinal image. The ROI of both the FFA and SLO images are divided into multiple overlapping blocks and each block is registered locally. A detailed explanation of the vessel extraction and global and local registrations is provided in the following.

3.5.1.1 Vessel extraction

The ‘exploratory Dijkstra forest’ algorithm [25] is utilized for vessel extraction and is an iterative algorithm. At each iteration, the best unvisited pixel in the vessel image is selected as a starting point for a dynamic-programming exploration of the unvisited part of the image. It results in creating a new tree in the growing forest of vessels. The forest growth is controlled with a threshold criterion.

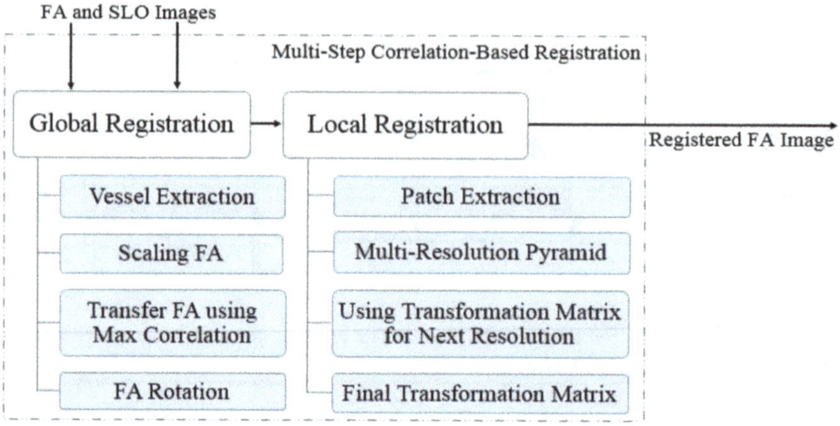

Figure 3.13. A diagram of the multi-step correlation-based registration technique in [24]. (Reproduced with permission [24]. Copyright 2019 IEEE.)

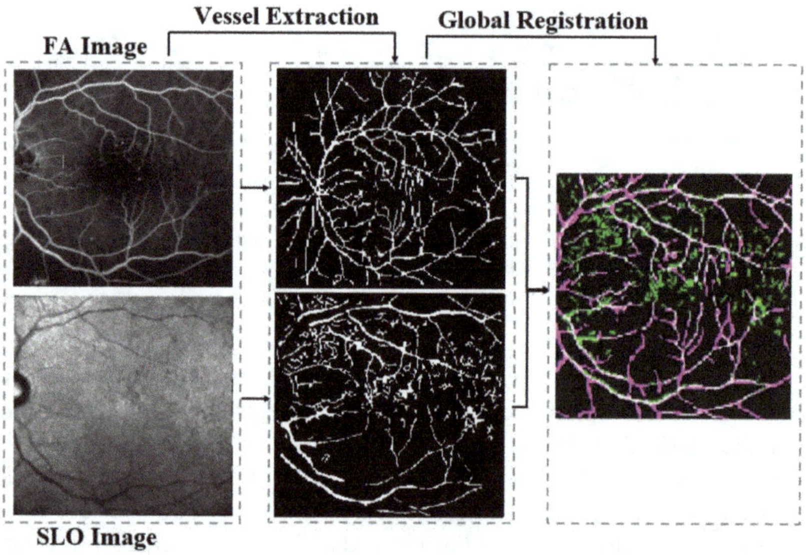

Figure 3.14. Vessel extraction and global registration results for a pair of SLO/FFA sample images [24]. (Reproduced with permission [24]. Copyright 2019 IEEE.)

3.5.1.2 Rigid or global registration

After the vessel extraction phase, an iterative multi-step correlation-based registration algorithm is performed on the vessel maps of the FFA and SLO binary images. In each iteration, the scale, shift or transfer, and rotation parameters are updated in three different steps and finally optimized at the end of the iterations. The result of vessel extraction along with three-step global registration for a sample image is depicted in figure 3.14.

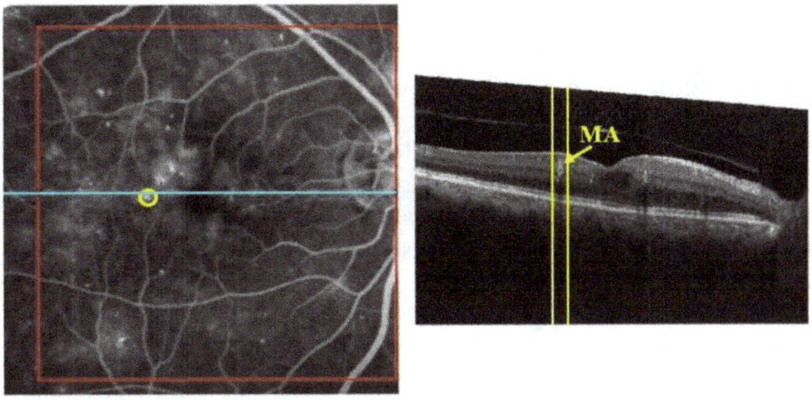

Figure 3.15. Left: labeled MA spot on the FFA image. Right: corresponding MA region on the OCT B-scan for a sample image [24]. (Reproduced with permission [24]. Copyright 2019 IEEE.)

3.5.1.3 Non-rigid or local registration

As global registration does not achieve perfect registration results in different regions of the images, a patch-based local registration is performed on some ROI (regions containing a microaneurysm (MA) labeled by experts) to improve the registration result for those parts [24]. The local processing is performed by a sliding window of 40×40 dimensions. The patch is centered at the same indexed pixels for both the FFA and SLO images. An intensity multi-resolution registration method is performed (utilizing MATLAB implementation of a method called *imregister*) on this patch. To achieve the best results, a multi-resolution decomposition technique with different scales is performed repeatedly on each patch to find the optimal registration parameters for all the pixels. The result of this step for a sample image is shown in figure 3.15.

The MA spot is highlighted by a yellow circle in the FFA image, and the corresponding MA region is highlighted by yellow bars in the OCT B-scan. You can see the MA as a hyperreflective structure between the yellow bars in the OCT B-scan. The accurate determination of MA regions on OCT B-scans represents the accurate and precise performance of the registration algorithm. Their overall reported empirical results are promising.

3.5.2 MA detection in OCT and FFA

3.5.2.1 MA detection in FFA

Diabetic retinopathy is a hazardous complication of diabetes that causes variations in blood vessels of the retina and in higher stages leads to blindness. A retinal MA is a very small area of blood sticking out from an artery or vein in the retina and is the first visible sign associated with DR [26]. Since MAs occurring in the macula threaten vision, early diagnosis of them can help to reduce the incidence of blindness. Furthermore, different stages of diabetic retinopathy can be analyzed and classified knowing the number of MAs and position of them relative to the macula. However, MAs are of low contrast and small size, about 25 to 100 microns, and it is hard to automatically detect these tiny lesions. The FFA is a more distinctive way to detect

MAs than color fundus images and the detection of these lesions from FFA is less complex. A new method for MA detection in FFA was demonstrated in [27] based on curvelet transform and morphological operations. In the first step of this method vessels were extracted from enhanced FFA image by a curvelet transform. In the second step extracted vessels were subtracted from original images. In the third step, morphological operations, dilation and erosion, were applied to the results of step two. Finally, the background of the image was mapped to zero and MAs were detected by thresholding. Sample results of different stages of this method are depicted in figure 3.16. This method was applied on 70 FFA images with different stages of DR and for each image the total number of MAs and the number of MAs in the FAZ were determined.

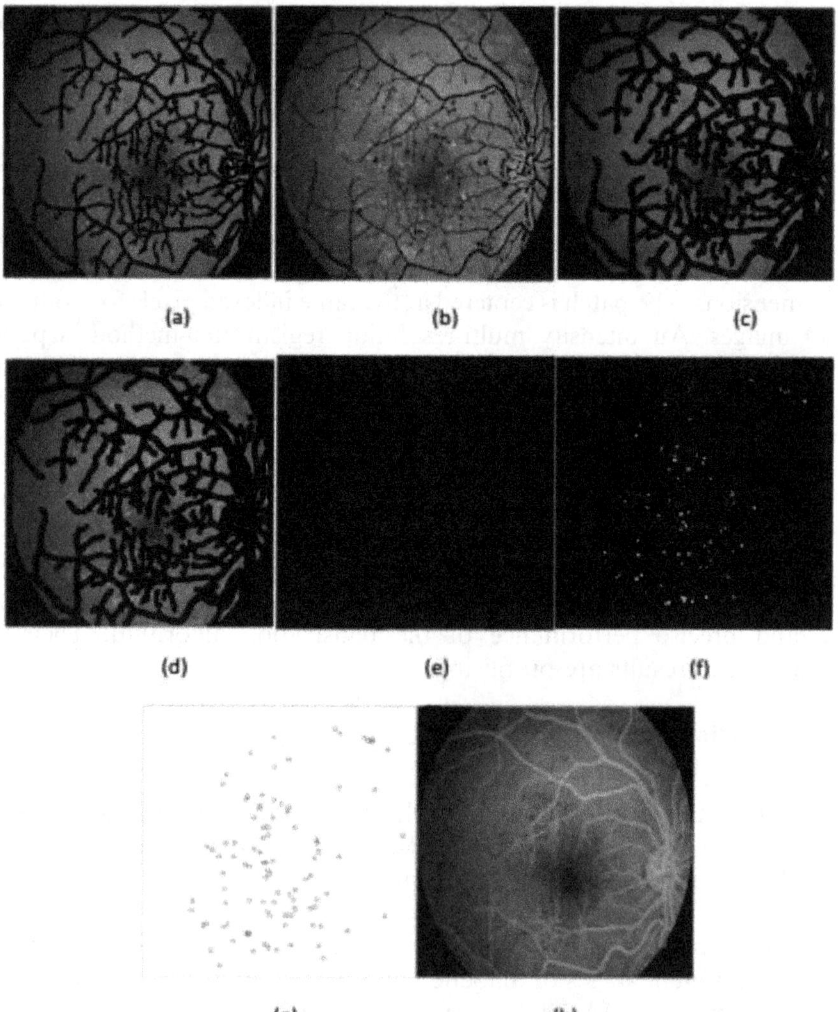

Figure 3.16. Sample results of MA detection in FFA from [26]. (Reproduced with permission from [26]. Copyright 2020 Elsevier.)

3.5.2.2 MA detection in OCT

Over the last few decades several clinical studies have focused on MA detection and analysis in OCT images of the retina of DR and diabetic macular edema (DME) patients. The clinical research studies focus on type, shape, features, location, and frequency of MAs in patients and finding possible correlations between the severity of the disease and the distribution of MAs in retinal OCT images.

Some clinical studies [28–30] have tried to describe the foci features residing in the retina of all DME patients and presented as well-demarcated hyperreflective patterns in the corresponding OCT B-scan images. The foci were present in all retinal layers. They also studied the shape and size of MAs in different retinal layers. Another group of studies [31–33] focused on studying the retinal layers containing MAs and tried to find patterns describing it. The third group of researchers [34–36] studied the correlation between OCT and FFA patterns in DR and DME patients. Horii *et al* [30] studied different types of ring signs indicating MAs and concluded that hyperreflective spots and cystoid spaces are more common in MA cases with incomplete ring signs. By leveraging and integrating all these findings, we present a comprehensive description of MA related features and properties as follows:

1. Capsular structure (i.e. ring sign).
2. Presence of shadow under the MA in outer layers.
3. Corresponding retinal layer.
4. Internal reflectivity.
5. Presence of cystoid spaces.
6. Lumen reflectivity of nearby cystoid spaces.
7. Presence of hyperreflective foci.
8. Adjacent hyperreflectivity around MAs.
9. Similar properties on consecutive B-scan images.

The capsular structure or ring sign may have three different states: complete, incomplete, or absent (see figure 3.17). In figure 3.17(a) the MA is surrounded by a slightly darker border which indicates the presence of a complete capsular structure (ring sign) around the MA. The dark border is partial in figure 3.17(b) which forms

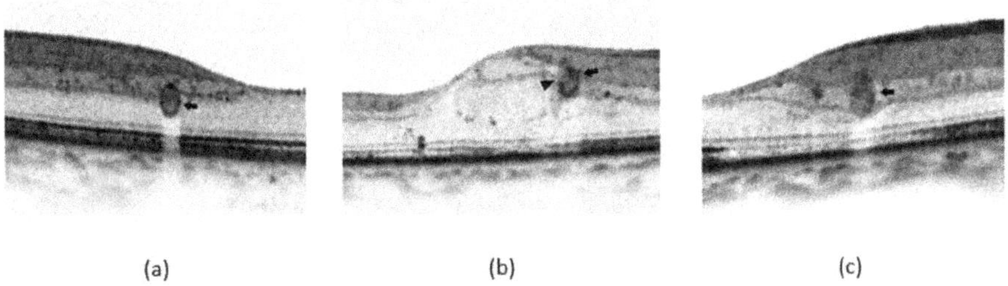

Figure 3.17. Different types of capsular structure or ring sign for MA. The black arrow indicates the MA. (a) Complete ring sign, (b) incomplete ring sign, and (c) absent ring sign [30]. (Reproduced with permission from [30]. Copyright 2010 Elsevier.)

an incomplete ring sign. The dark border is missing in figure 3.17(c) which indicates the absence of a ring sign around the MA.

In some studies, the presence of shadows underneath MAs was reported, hence the presence or absence of shadow under MAs is an important feature to consider. The shadow in the OCT image is presented as a vertical column of discoloration (lighter or darker color depending the type of capturing device) that is placed right underneath the MA's body. In figure 3.14(a) the shadow under the MA is visible as a brighter vertical column passing through the inner retinal layers.

The MA residing retinal layers differs in different studies, but the most frequent residing layers are the retinal nerve fiber layer (RNFL)/ganglion cell layer (GCL), inner nuclear layer (INL)/inner plexiform layer (IPL), and outer nuclear layer (ONL)/outer plexiform layer (OPL). The internal reflectivity could have three different levels: hyper-, moderate-, or hypo-reflectivity. The cystoid spaces could be present, nearby the MAs, adjacent to the MAs, or protruding into the MAs. The lumen reflectivity of cystoid spaces could be homogeneous or heterogeneous (see figure 3.18).

The hyperreflective foci could be absent or present within the interior or posterior vessels walls of MAs. The presence or absence of adjacent hyperreflectivity around MAs in an important feature as well (see figure 3.19).

The foci are shown with red and black arrows. There is also a green circle around a dense group of foci emphasizing their distribution density. The foci are residing in different retinal layers and either near cystoid spaces or far away from them. Finally, working with B-scan images provides the possibility of studying and comparing consecutive B-scans. By comparing the previous and next B-scan of the current B-scan image and checking the presence of similar properties in them, one can have a sense of the volume and size of the MA and its visible effects on several consecutive B-scan images. In the case that the severity of progression of the disease is of interest, there is also a direct relationship between the density and distribution of MAs with the severity of disease progression in DME and DR patients.

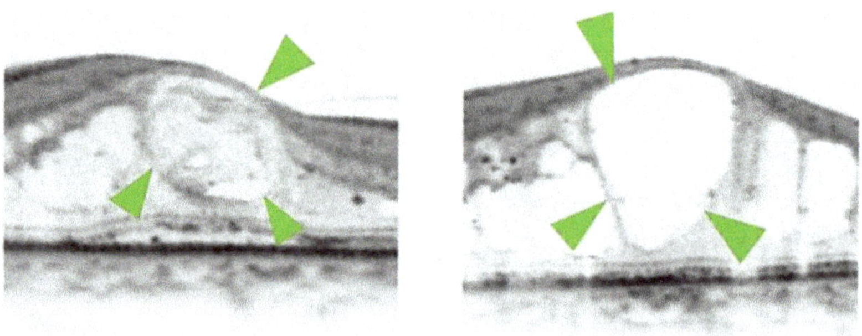

(a) (b)

Figure 3.18. Lumen reflectivity in cystoid spaces. (a) High heterogeneous reflectivity and (b) low homogeneous reflectivity. (Reproduced with permission from [37]. Hindawi Open Access.)

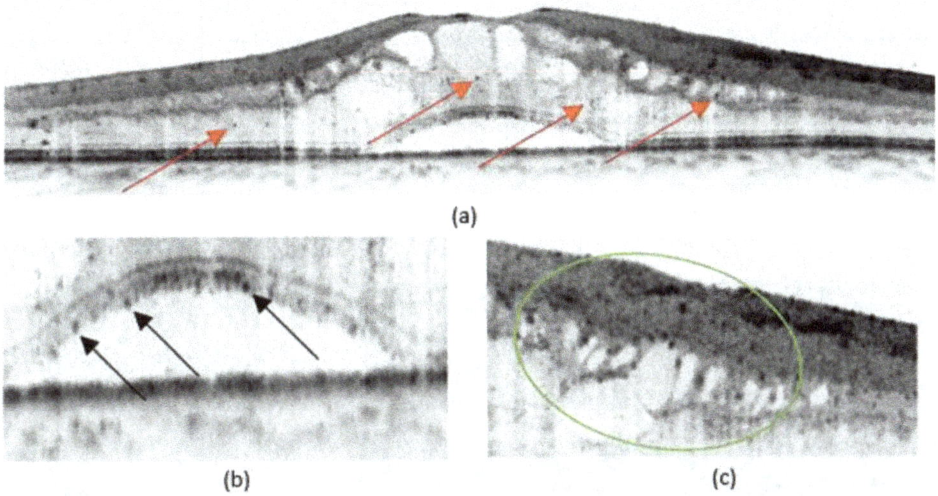

Figure 3.19. Hyperreflective foci are shown in all three OCT images.

Even though there is a great body of clinical work focused on studying MAs and their corresponding features, progress in computational studies regarding the automatic detection of MAs in OCT images has been slower. The lack of sufficiently large datasets from different OCT capturing devices, the lack of standardized and unified imaging and post-processing protocols between devices, and limited graphics processing unit (GPU) capabilities for exploiting three-dimensional features are the main difficulties in the automatic analysis of OCT images [38]. By designing an automatic or a semi-automatic MA detection system, the described clinical features could be analyzed and validated which would provide great help to clinician experts in detecting and studying MAs in DR and DME patients and hence accelerating the treatment process. Different image analysis techniques and methods could be utilized in designing such a system. A straightforward and relatively simple algorithm with several processing steps is described in the following.

To extract MAs and hyperreflective foci from the OCT B-scan image, a multi-step process is performed. The B-scan image needs to be normalized in the range of [0–1] to be fed to the extraction algorithm. Referring to different studies, MA could be residing at several different retinal layers. However, the common factor in all the studies is that no MA is residing below the INL and above the RPE layers. Therefore, detecting these two borders and excluding them from the region of interest (BW image containing all the hyperreflective features in the image) helps in ultimate detection of MAs and foci. To achieve MA and foci detection, a three-step algorithm is performed. This algorithm consists of Otsu thresholding, illumination equalization (contrast enhancement), and border detection using a general graph method. The labeled B-scan image with highlighted MAs and foci mapped on it is the output of the algorithm. The flowchart of the algorithm is depicted in figure 3.20.

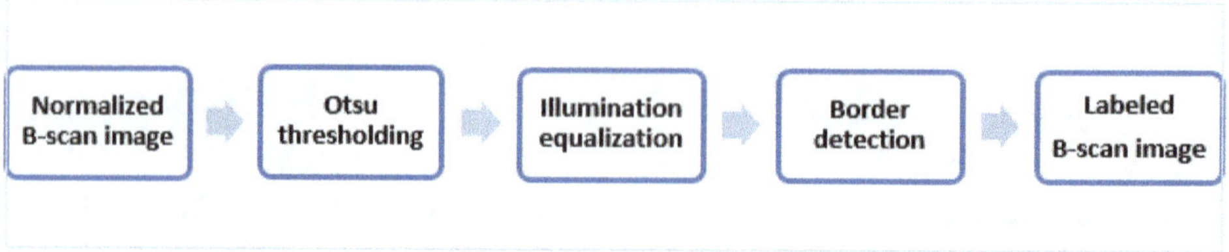

Figure 3.20. MA and foci detection algorithm flowchart.

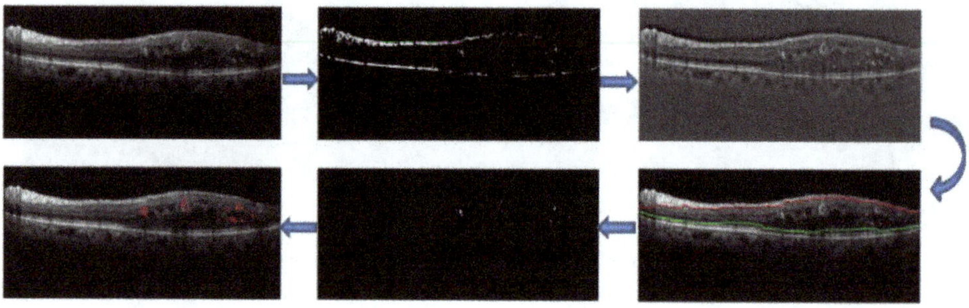

Figure 3.21. MA and foci detection algorithm for a B-scan of a DR patient. One MA with a complete ring sign and several foci are residing in the middle layers on the image. A major detachment is also visible between the retinal layers.

This algorithm is applied on several OCT B-scans of the Heidelberg device. Some of results are promising and some need more improvement. The result of the algorithm for a sample B-scan image is depicted in figure 3.21. The algorithm successfully marked the residing MA and foci in the B-scan image. The red dots are the spots detected by the algorithm.

3.6 Conclusion

In this chapter we discussed the application of enface image registration/alignment to introduce new ocular imaging biomarkers. Ocular biomarkers provide an objective, measurable method of evaluating the progress of various eye diseases from diagnosis to treatment. Desirable biomarkers can be extracted from various modalities of enface retinal imaging such as fundus images, SLO, and FFA images. Although features from the landmarks of retina such as retinal vessels, ONH, and FAZ can be used as guiding markers to diagnose retinal disorders, the most effective biomarkers can be achieved by registration and alignment of different modalities of enface imaging. For instance, the asymmetric pattern of the vascular map in both eyes, symmetry analysis of morphology, and the thickness of retinal layers in both eyes, and the local cup-to-disc ratio of the optic nerve head are efficient biomarkers which can be determined after registration of the OCT images to aligned fundus images. These biomarkers can be used in the diagnosis of various retinal disorders. Other biomarkers that are widely used to analyze different stages of diabetic retinopathy are microaneurysms which can either be determined from FFA and

OCT images or during global or local registration of these images using SLO images. We reviewed some of the works that have been done so far to find such biomarkers using modalities of enface imaging. However, since non-invasive ocular biomarkers derived from enface imaging of retina are a potential tool for early diagnosis of various diseases, more studies must be done to find and validate new biomarkers to contribute decisively to the process of disease diagnosis and treatment.

References

[1] Remington L A and Goodwin D 2011 *Clinical Anatomy of the Visual System E-Book* (Amsterdam: Elsevier Health Sciences)

[2] Kafieh R *et al* 2019 Automatic multifaceted Matlab package for analysis of ocular images (AMPAO) *SoftwareX* **10** 100339

[3] Alipour S H M, Rabbani H and Akhlaghi M 2014 A new combined method based on curvelet transform and morphological operators for automatic detection of foveal avascular zone *Signal, Image Video Process.* **8** 205–22

[4] Rabbani H, Kafieh R and Amini Z 2016 Optical coherence tomography image analysis *Wiley Encyclopedia of Electrical and Electronics Engineering* ed J G Webster (New York: Wiley)

[5] Abràmoff M D, Garvin M K and Sonka M 2010 Retinal imaging and image analysis *IEEE Rev. Biomed. Eng.* **3** 169–208

[6] Sharp P F and Manivannan A 1997 The scanning laser ophthalmoscope *Phys. Med. Biol.* **42** 951

[7] Wong D 1977 Fundus photography and fluorescein angiography *J. Biol. Photogr. Assoc.* **45** 104

[8] Oshima Y 2010 Fundus photography and fluorescein angiography *Nihon Rinsho* **68** 262

[9] Esmaeili M, Rabbani H, Mehri A and Dehghani A 2009 Extraction of retinal blood vessels by curvelet transform *2009 16th IEEE Int. Conf. on Image Processing (ICIP)* pp 3353–6

[10] Golabbakhsh M, Rabbani H and Esmaeili M 2012 Detection and registration of vessels of fundus and OCT images using curevelet analysis *12th IEEE Int. Conf. on Bioinformatics and Bioengineering* pp 594–7

[11] Niemeijer M, Staal J, van Ginneken B, Loog M and Abramoff M D 2004 Comparative study of retinal vessel segmentation methods on a new publicly available database *Proc. SPIE* **5370** 648–56

[12] Soltanipour A, Sadri S, Rabbani H, Akhlaghi M and Doost-Hosseini A 2013 Vessel centerlines extraction from fundus fluorescein angiogram based on Hessian analysis of directional curvelet subbands *IEEE Int. Conf. on Acoustics, Speech and Signal Processing* pp 1070–4

[13] Esmaeili M, Rabbani H, Dehnavi A M and Dehghani A 2012 Automatic detection of exudates and optic disk in retinal images using curvelet transform *IET Image Process.* **6** 1005–13

[14] Jamshidi M, Rabbani H, Amini Z, Kafieh R, Ommani A and Lakshminarayanan V 2016 Automatic detection of the optic disc of the retina: a fast method *J. Med. Signals Sens.* **6** 57

[15] Miri M, Amini Z, Rabbani H and Kafieh R 2017 A comprehensive study of retinal vessel classification methods in fundus images *J. Med. Signals Sens.* **7** 59

[16] Mahmudi T, Kafieh R, Rabbani H, Mehri A and Akhlagi M 2015 Asymmetry evaluation of fundus images in right and left eyes using radon transform and fractal analysis *IEEE Int. Conf. on Image Processing* pp 163–7

[17] Kafieh R, Danesh H, Rabbani H, Abramoff M and Sonka M 2013 Vessel segmentation in images of optical coherence tomography using shadow information and thickening of retinal nerve fiber layer *IEEE Int. Conf. on Acoustics, Speech and Signal Processing* pp 1075–9

[18] Mahmudi T, Kafieh R, Rabbani H, Dehnavi A M and Akhlaghi M R Evaluation of asymmetry in right and left eyes of normal individuals using extracted features from optical coherence tomography and fundus images *J. Med. Signals Sens.* **11** 12–23

[19] Kafieh R, Rabbani H, Abramoff M D and Sonka M 2013 Intra-retinal layer segmentation of 3D optical coherence tomography using coarse grained diffusion map *Med. Image Anal.* **17** 907–28

[20] Kafieh R, Rabbani H, Hajizadeh F, Abramoff M D and Sonka M 2015 Thickness mapping of eleven retinal layers segmented using the diffusion maps method in normal eyes *J. Ophthalmol.* **2015** 259123

[21] Kafieh R, Rabbani H, Abramoff M D and Sonka M 2013 Curvature correction of retinal OCTs using graph-based geometry detection *Phys. Med. Biol.* **58** 2925

[22] Mokhtari M, Rabbani H, Mehri-Dehnavi A and Kafieh R 2017 Exact localization of breakpoints of retinal pigment epithelium in optical coherence tomography of optic nerve head *39th Annual Int. Conf. of the IEEE Engineering in Medicine and Biology Society* pp 1505–8

[23] Mokhtari M *et al* 2019 Local comparison of cup to disc ratio in right and left eyes based on fusion of color fundus images and OCT B-scans *Inf. Fusion* **51** 30–41

[24] Kamasi Z G, Mokhtari M and Rabbani H 2017 Non-rigid registration of fluorescein angiography and optical coherence tomography via scanning laser ophthalmoscope imaging *39th Annual Int. Conf. of the IEEE Engineering in Medicine and Biology Society* pp 4415–8

[25] Rabbani H, Allingham M J, Mettu P S, Cousins S W and Farsiu S 2015 Fully automatic segmentation of fluorescein leakage in subjects with diabetic macular edema *Invest. Ophthalmol. Vis. Sci.* **56** 1482–92

[26] Amini Z, Kafieh R, Mousavi E and Rabbani H 2020 Diabetic retinopathy detection in ocular imaging by dictionary learning *Diabetes and Fundus OCT* (Amsterdam: Elsevier) p 343

[27] Alipour S H M and Rabbani H 2013 Automatic detection of micro-aneurysms in retinal images based on curvelet transform and morphological operations *Proc. SPIE* **vol 8856** 88561W

[28] Bolz M *et al* 2009 Optical coherence tomographic hyperreflective foci: a morphologic sign of lipid extravasation in diabetic macular edema *Ophthalmology* **116** 914–20

[29] Hasegawa N, Nozaki M, Takase N, Yoshida M and Ogura Y 2016 New insights into microaneurysms in the deep capillary plexus detected by optical coherence tomography angiography in diabetic macular edema *Investig. Ophthalmol. Vis. Sci.* **57** OCT348–55

[30] Horii T, Murakami T, Nishijima K, Sakamoto A, Ota M and Yoshimura N 2010 Optical coherence tomographic characteristics of microaneurysms in diabetic retinopathy *Am. J. Ophthalmol.* **150** 840–8.e1

[31] Lee S N *et al* 2013 Characterization of microaneurysm closure after focal laser photocoagulation in diabetic macular edema *Am. J. Ophthalmol.* **155** 905–12

[32] Wang H *et al* 2012 Characterization of diabetic microaneurysms by simultaneous fluorescein angiography and spectral-domain optical coherence tomography *Am. J. Ophthalmol.* **153** 861–7.e1

[33] Yamada Y, Suzuma K, Fujikawa A, Kumagami T and Kitaoka T 2013 Imaging of laser-photocoagulated diabetic microaneurysm with spectral domain optical coherence tomography *Retina* **33** 726–31

[34] Yeung L, Lima V C, Garcia P, Landa G and Rosen R B 2009 Correlation between spectral domain optical coherence tomography findings and fluorescein angiography patterns in diabetic macular edema *Ophthalmology* **116** 1158–67

[35] Soliman W, Sander B, Hasler P W and Larsen M 2008 Correlation between intraretinal changes in diabetic macular oedema seen in fluorescein angiography and optical coherence tomography *Acta Ophthalmol.* **86** 34–9

[36] Tan B *et al* 2020 Quantitative microvascular analysis with wide-field optical coherence tomography angiography in eyes with diabetic retinopathy *JAMA Netw. Open* **3** e1919469

[37] Murakami T and Yoshimura N 2013 Structural changes in individual retinal layers in diabetic macular edema *J. Diabetes Res.* **2013** 920713

[38] Yanagihara R T, Lee C S, Shu D, Ting W and Lee A Y 2020 Methodological challenges of deep learning in optical coherence tomography for retinal diseases : a review *Assoc. Res. Vis. Ophthalmol.* **9** 11

IOP Publishing

Photo Acoustic and Optical Coherence Tomography Imaging, Volume 2
Fundus imaging for the retina
Ayman El-Baz and Jasjit S Suri

Chapter 4

Existing techniques used for retinal image analysis in the automated detection and prediction of AMD

Samina Khalid, Usman Akram, Tehmina Sheryar and Farah Karim

4.1 Introduction

Applications of image processing techniques have been developing in the field of medical sciences in the recent era. These image processing techniques and processes are helpful in the timely diagnosis of different diseases. Medical imaging consists of image acquisition techniques that are used non-invasively to produce images of the internal features of human body parts that are invisible to the naked eye, for example x-rays and magnetic resonance imaging (MRI) images. Different techniques have been used for the purpose of measuring and recording medical information, but techniques such as electrocardiography (EKG), electroencephalography (EEG), and magnetoencephalography (MEG) are not primarily designed to generate images. Currently, computer aided diagnosis (CAD) systems are widely used to assist physicians in the interpretation of medical images. These CAD systems yield useful information which can be used by physicians and medical specialists to analyze and assess information comprehensively in a short period of time. In the present era, different types of MRI, fundoscopic, and x-ray images (scans) are being used in CAD systems, as shown in figure 4.1.

In the field of medicine, x-ray images were initially used by radiologists for the diagnosis and/or treatment of bone fractures. The concept of a CAD system was introduced primarily to assist physicians in the early diagnosis of breast cancer on mammograms (x-rays of the chest). Currently, CAD systems have become an essential part of the routine clinical work for the diagnosis of various diseases. In the field of ophthalmology, the use of a CAD system also helps ophthalmologists in the

doi:10.1088/978-0-7503-2056-6ch4

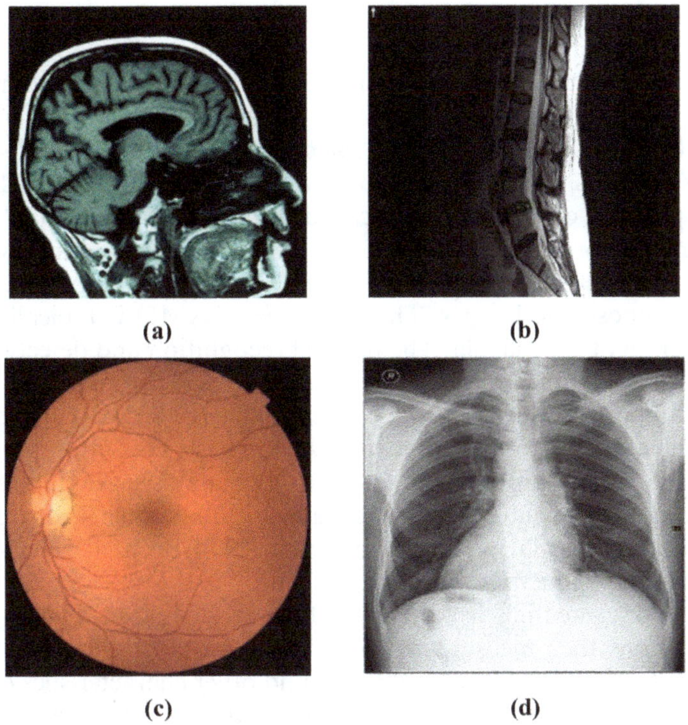

Figure 4.1. Digital images used by CAD systems for the diagnosis of different diseases. (a) MRI scan of the head. (b) MRI scan of the spinal cord. (c) Fundoscopic retinal image. (d) Chest x-rays.

Figure 4.2. Images: (a) normal eye and (b) AMD affecting the eye.

timely diagnosis and mass-screening of various retinal diseases such as diabetic retinopathy, glaucoma, and age-related macular degeneration (AMD), etc. AMD is the leading cause of blindness worldwide in the elderly age group. Generally, this macular disorder is known as the third leading cause of blindness globally after cataracts and glaucoma. The AMD is a bilateral ocular condition that affects the central area of the retina known as the macula and causes problems with central vision. Figure 4.2 shows the fine detail of an AMD affected macula eye compared to a healthy macula eye. Although the macula comprises only 4% of the retinal area, it is responsible for the majority of useful photonic vision.

AMD is the main cause of blindness in the elderly in developed countries such as Australia, the UK, and the USA. The estimated annual cost burden from AMD in the USA is 30 billion USD or about 0.3% of the gross domestic production. The occurrence of AMD is expected to double over the next 25 years. AMD originates due to the deposition of bright lesions called drusens in the macular region of the retina. Drusens are formed at the retinal level and can affect central eyesight. A lot of research has been done in the field of medical care for the automated detection of drusens from fundoscopic images. The diagnosis of AMD is typically undertaken through inspection of the macula. The manual recognition and detection of drusens from retinal images is time-consuming and expensive. Moreover, this technique is subjective and reproducibility is also a concern. To save on workload and facilitate large-scale clinical use, it is important to have a precise, cost-effective, and efficient system to detect drusens automatically for AMD diagnosis. AMD is characterized by the appearance of the sub-retinal deposits medically known as drusens as well as abnormalities of the retinal pigment epithelium (RPE) retinal layer consisting of depigmentation and/or increased retinal pigment.

Identifying and segmenting drusens on a retinal image is central in the classification of AMD. Therefore, their measurement and quantification are important in research and various clinical studies. The size, density, and other characteristics of drusens are used to determine the classification of the stage of the disease to serve as a basis for establishing a more refined phenotype of AMD. There are innumerable techniques proposed in the literature that ophthalmologists are using to detect retinal disorders from human retina. A number of algorithms have been proposed for the identification of drusen patterns from the macular region of the retina and classification of AMD from retinal images. Ophthalmoscopy, fundoscopic photography (FP), and fluorescein angiography (FA) are the imaging modalities with applications of image processing techniques that are used extensively in the field of ophthalmology for the automated diagnosis of AMD. A lot of useful techniques have been developed for the automated diagnosis of eye diseases and retinal abnormalities using pattern recognition and image processing. However, most of these techniques have been developed for fundoscopic image analysis.

The drawback of fundoscopic images is that many diseases such as diabetic retinopathy and other AMD diseases are misdiagnosed in their early stages. This is because fundoscopic images may not show any sign of the disease present at an early stage. In this case, the use of the optical coherence tomography (OCT) imaging technique is of significant importance for early detection of retinal diseases. One of the major advantages of the OCT imaging system is that it can provide early detection of all retinal disorders compared to other imaging techniques. OCT imaging can also give an objective evaluation of macular and ocular pathology as it shows a cross-sectional region of the macula and optic disc part of the retina. Another advantage of OCT imaging is that it acquires fundoscopic images as well. For that reason, by using a single camera we can obtain both images for the same patient at the same time. This can add in a more detailed severity analysis of retinal diseases for ophthalmologists to further diagnose the disease. Medical image processing is an emerging field of research that is important and very helpful for the development of CAD systems. These systems are found helpful in the timely diagnosis of retinal disorders before these disorders cause irreversible blindness. In this chapter, a detailed literature analysis is performed in the context of the image processing techniques used to date for the automated diagnosis of AMD retinal disorder from fundoscopic and OCT images. Drusen presence is known as the hallmark of AMD and can be identified as a white and yellowish type of bright lesion in fundoscopic images [1], whereas in OCT images these drusens can be seen as bumps in the RPE layer of the retina [2]. The number and size of drusen occurrence in the macular region can be further used to distinguish different stages of the dry (non-exudative, non-neovascular) form of AMD [3]. Different image pre-processing techniques, i.e. image enhancement, image segmentation, and image reconstruction, can be used for differentiating retinal abnormalities from fundoscopic well as from OCT images.

4.2 Existing automated diagnosis methods used for AMD

Various methods for automated diagnosis of AMD have been proposed in the literature [4–8]. Previously proposed methods that have great implications for the automated diagnosis of AMD are summarized in this section. These methods consist of pre-processing, retinal layer and drusen detection and segmentation, and AMD diagnosis modules. In the coming sections, the different image processing techniques used for each module are discussed in detail.

4.2.1 Pre-processing techniques

The pre-processing step is done after acquiring an image to remove/reduce noise in order to enhance its quality for further processing. A pre-processing step is crucial to highlight different attributes that can be further used during the object segmentation process. Basically, there are two techniques that are most extensively used in the segmentation process, known as contrast enhancement and filtering. In [5, 6, 9–11] histogram-based approaches have been used for color and contrast enhancement of fundoscopic retinal images. A median filter in [7], anisotropic diffusion in [8], and hue, saturation, and value (HSV) channels in [10] have been used for image de-noising. For image segmentation, thresholding techniques have been found to be very significant and are widely used because of their simplicity and effectiveness [12, 13]. Thresholding extraction is performed to segment out pixels that represent some objects in an image foreground. This has been achieved with the help of optimal thresholding by converting grayscale images into binary images to extract foreground pixels from the images which represent some object [14].

4.2.2 Retinal component extraction from fundoscopic images

For the automated diagnosis of AMD from fundoscopic images, first we need to detect drusens from the macular part of the retina. These drusens occur in the macular surface part of the retina. On the basis of the detection of these drusens, further progression of AMD disorder can be determined. Different techniques for drusen detection from fundoscopic images presented in the existing literature are documented in the next section.

4.2.2.1 Drusen detection

Drusen detection from a fundoscopic retinal image and its segmentation are known as crucial steps to classify AMD. Drusen measurement and quantification on the basis of their size, density, shape, and area help to determine different stages of dry and wet AMD. Therefore, these two steps are central in research and various clinical studies [3, 15]. The detection, quantification, and measurement of drusens help to determine the severity of the retinal disorder and serve as an important process in the development of an automated diagnostic system as well as for the evaluation of the development of disease. In [7] Liang *et al* presented a method for drusen detection from fundoscopic retinal images. Blood vessels were first removed because these blood vessels could create confusion while performing macula detection due to similar intensities. The authors utilized green and red channels and performed edge detection to determine high contrast locations, representing blood vessels. After the detection of edges, the outcomes of the used channels were added to generate a binary mask to check whether an individual pixel belongs to the vessel or non-vessel part of the macula. Morphological dilation with a square pixel kernel of size 3×3 was applied for detecting regions smoothness. The authors suggested that the required drusenoid region can be spotted when edge transformation is performed. An area with 300×300 pixels as a region of interest (ROI) was characterized for determining macula location by keeping the center around the focal piece of the retinal fundoscopic image. For the extraction of the lowest value of intensity from a grayscale image, a pixel search was performed within the ROI. These low intensity searched pixels were further used as seeds for identifying groups of pixels with similar values. From these low intensity pixel groups only the region that consisted of the large pixel group was identified as the macula. After macular region identification, further detection of the macula based on grading overlay was performed and drusen detection was accomplished from a region within three disc diameters (3DD) of the macula. In drusen detection, the presence of choroidal vessels in the green channel was prone to errors. Therefore, it was important to suppress choroidal vessels by using the red channel before drusen detection. Based on RGB and HSV channel characterization, drusen spots were detected. The shortcoming of their proposed system was that they only considered drusens present near the outer circle of the grading overlay while drusens presented anywhere in macular region could cause the AMD disorder.

A comparative analysis of fundoscopic images with FA and spectral domain OCT (SD-OCT) images regarding their sensitivity and specificity for AMD detection, choroidal neo-vascularization (CNV), and CNV activity is presented by Nils *et al* in [16]. After analyzing the three imaging technologies using images of 120 patients' eyes, the authors concluded that fundoscopic imagery is helpful for identifying drusen and pigmentary changes, but this imaging technology may be failing to detect CNV lesions and activity. In [17] Brandon *et al* utilized a multi-level procedure, beginning from the pixel level classification and after that to the area level, region level, and finally considering image level classification, which permitted the most minimal levels of classification to be tuned to detect even the faintest regions to discriminate as drusens, relying on the higher levels of classification to use an ever-broadening context to refine the segmentation. Using a dataset of 119 fundoscopic images, 87% accuracy was achieved by their proposed system. A method based on a grading system for drusen measurement was proposed in [18] to measure drusen size. For the detection and extraction of drusens, an image processing technique based on a statistical method was shown in [19] to grade AMD progression from a longitudinal image dataset. As an initial step, fundoscopic images were normalized for controlling change caused due to illumination. Afterward, categorization of retinal parts was performed to choose candidate drusen pixels. After that drusen detection was performed for the characterization of every pixel either as a part of a drusen structure or not. In the last step, morphological operations were performed to remove minor or large drusen holes. The only limitation of their proposed system was that hard drusens were not considered in the diagnosis of early AMD, but these structures actually cause early AMD disorder.

A positive correlation between the risk of progression of wet AMD with respect to drusen size has been found in the literature [3, 19, 20]. For the dry form of AMD, color fundoscopic image evaluation has been performed. The size and area of drusens have been compared with a standardized set of circles for drusen size estimation. Localizing drusens from fundoscopic images is found challenging because of the changing background of pigments of the retinal macular surface. This study found that it was important to understand how the drusenoid area can be identified in OCT and FA images with respect to drusen size and area as identified using color fundoscopic images. Sbeh *et al* in [21] proposed a strategy for the extraction of angiographic eyes from fundoscopic images. Drusen extraction was performed at the retina level by utilizing a versatile calculation dependent on numerical morphology transforms. Sbeh *et al* in [22] correspondingly exhibited a novel segmentation technique. In their proposed transformation method they utilized finding a new class of local maxima from the input retinal image. In [23] Rapantzikos *et al* proposed a system based on a new segmentation technique. The histogram-based adaptive local thresholding (HALT) technique was applied to distinguish drusens in fundoscopic images by extracting the useful evidence without being influenced by the presence of other pathological structures. Aside from this, distinctive fuzzy logic based strategies [24] and surface textural based techniques [25, 26] have been proposed for drusen localization and extraction from fundoscopic images. In [27], based on a differential algorithm, Rani *et al* proposed a method for the automated diagnosis of macular degeneration using fundoscopic retinal images. In their proposed system, first the optic disc was detected and segmented out from the input image. Then, using adaptive histogram equalization and thresholding, drusen detection was performed. The performance of the proposed system was tested using the STARE dataset. Their proposed system achieved 82%, 85.7%, and 66.7%, accuracy, sensitivity, and specificity, respectively.

In [28] the authors proposed an automated drusen segmentation method for the diagnosis of AMD screening from retinal fundoscopic images. Their proposed method was based on Reny's entropy threshold. The authors performed statistical analysis based on results generated by manual segmentation and proposed an automated drusen detection system. A publicly available STARE dataset was used for testing and their proposed method achieved 93.37% sensitivity in classifying true positive cases. In [29] Mohaimin *et al* proposed a novel method based on a combination of color and object boundaries for accurate detection of drusens. Color normalization, a method based on non-uniform illumination correction, has been proposed to minimize inter-subject unpredictability. From the analysis using ARIA, a publicly available fundoscopic image dataset, their proposed system achieved an overall accuracy of 96.62%.

4.2.2.2 AMD lesion detection

In the process of an automated AMD diagnostic system, the automated detection of lesions and their classification from retinal images is the last stage. There are various techniques developed for the diagnosis of AMD. For automated mass-screening of retinal images in [9], the classification of pre-labelled (AMD +ve and −ve) images, a histogram-based approach has been used. A new input retinal image is then classified to a histogram that has the best wrapping path. To analyse the effectiveness of using a colored channel histogram and Hui saturation intensity (HSI) histogram an experiment was performed on 144 hand-labeled fondoscopic images; 86 of these images were AMD affected and the rest were normal images. The HSI channel achieved better results in term of classification accuracy with 74% recorded by the saturation component, which shows the ability of HSI in identifying patterns through the colors of the image. Representative patterns of retinal images with and without drusens were extracted in the form of a histogram. Labelled exemplary histograms were then stored in a case-base. Unseen examples were then classified by comparison with this case-base and analysed for drusens using a dynamic time wrapping comparison process.

In [6] Jayanthi *et al* proposed a framework that helps in diagnosing human retinal diseases using an auto-associative neural network based classifier. In their study, they analyzed how texture-based features can help to identify the spread of disease in the retina. Different methods have been discussed with the purpose that these methods can be used to detect features, classify, and diagnose the type of retinal disease at an early stage. For early diagnosis of AMD, an automated method for drusen detection and quantification from fundoscopic images has been proposed in [18]. Earlier automated methods that have been proposed in the literature could detect drusens without quantifying its area, but identification and quantification of drusens are important for the classification of different stages of AMD, and there exist different size of drusens, i.e. soft and hard drusens, and soft indistinct and soft distinct drusens. Different size drusens can help further to detect whether it is an early, intermediate, or advanced stage of AMD. An automated grading and AMD classification method [20] has been developed in which a standard method for drusen measurement has been proposed. According to this standard, drusens which have a diameter of up to

63 microns are classified as hard drusens (early stage), drusens with a size between 63 and 125 microns belong to the intermediate stage, and soft drusen that are larger than 125 microns belong to the advanced stage of dry AMD. Further classes of drusens were also defined in this study based on drusen density, i.e. drusen with uniform density and a sharp margin, classified as soft distinct drusens, and drusens with ambiguous edges, known as the soft indistinct drusens class. Therefore, this automated method can be used for the detection, classification, and quantification of drusens into different sub-types of an early AMD diagnostic system. The Age-Related Eye Disease Study (AREDS) identified classes of early AMD according to the scale of severity: no AMD, early stage, intermediate stage, and advanced stage. In these four types of classes, each successive class type represents a higher risk of developing the advanced stage of AMD. This automated diagnosis system attained sensitivities of about 74.94% and 81.17%, respectively, in identifying intermediate stage and advanced stage drusens. A convolutional neural network (CNN) based algorithm was used in [30] to extract hemorrhages surrounding the macula. These hemorrhages affect the retina because of wet AMD. Their proposed algorithm was used to isolate the macula, hemorrhages, and drusens in retinal fundoscopic images. Histogram equalization was performed for contrast enhancement. The ROI was extracted by applying *K*-means clustering. Later, a cellular neural network classifier with a combination of erosion and dilation was performed for further processing. This technique was unable to identify properly those drusens which were very thin. However, this method has the potential for use in longitudinal studies.

In [31] Zheng *et al* proposed a system for automated assessment of the risk of AMD development. This system incorporates learning based drusen detection. Their study also includes analysis of fundoscopic imaging techniques used for de-noising, image enhancement, and color transfer. Validation of this automated system was performed by comparing the results produced by this system with those obtained via manual drusen segmentation. This proposed automated system outperformed clinical studies. An automated approach for discrimination of two bright lesions, exudates and drusens, presented in [32] consists of a two-stage process. Bright pixels from fundoscopic images were detected in pre-processing. These identified bright regions were then categorized as exudates or drusens using an support vector machine (SVM) classifier. Ophthalmologists find it difficult to differentiate between drusens and exudates because of their similarity in appearance on the retinal surface. Feature extraction was performed pixel-wise to detect bright lesions, and also region-wise for final differentiation of drusens from exudates. The system was found to be helpful for an ophthalmologist for the desired automated differentiation with an overall accuracy of 92%, with 92% and 90% specificity and sensitivity, respectively. In [33] Phan *et al* proposed an automated screening and grading system for AMD using textural pattern analysis from fundoscopic images. In their proposed method, AMD was characterized by considering textural and color features. To test a feature's relevance, the AREDS protocol was followed in their proposed method. To classify images into AMD and normal images, the SVM and random forest classifiers were used. A dataset of 279 fundoscopic images was used to check performance. The results of the experiment show that local binary pattern

(LBP) textural features are very relevant to give information about AMD classification. Their method achieves good performance with an area under the curve receiving operating characteristics (ROC) between 0.739 and 0.874 for screening and for grading 0.469 and 0.685. In this automated system drusens present in the peripheries of the retina were also considered instead of drusens present in the macular region only, as drusens present in different parts of the retina can cause different retinal disorders and drusens present only in the macular region can cause AMD. In [46] Kumari *et al* proposed an automated method for drusen detection for the diagnosis of AMD from fundoscopic images. The focus of their analysis was the quantification of drusens based on correction of non-uniform based illumination by performing region based statistical analysis. They used enhanced local intensities, noise removal, and segmentation of images using the OTSU algorithm in combination with a morphological operation, this information was further utilized to detect the area and edge of drusens. For the quantification of the drusen area, pixel based feature extraction was also performed. Using two publicly available datasets, STARE and ARIA, their proposed method achieved good accuracy in the identification and quantification of drusenoid areas.

An automated screening system for the classification of dry and wet AMD cases using the pyramid of histograms of orientation gradients (PHOG) and non-linear features was proposed in [34] by Rajendra Acharya *et al.* A dataset containing 945 fundoscopic images, including 404 normal, 517 dry AMD, and 24 wet AMD, was employed to test the performance of their proposed system. In their proposed approach they applied the PHOG method to extract discriminations in the pixels of fundoscopic images. Different non-linear features were further extracted from the PHOG descriptor. The authors used an adaptive synthetic sampling (ADASYN) method for balancing the number of fundoscopic images in all three classes (normal, dry, and wet). For more appropriate feature set selection, two feature selection methods, ant colony optimization genetic algorithm (AGO-GA) and particle swarm optimization (PSO), were used for achieving good performance. After feature selection, highly significant features for classification were further determined using analysis of variance (ANOVA). The proposed method achieved an overall accuracy of 85.1%, with 87.2% and 80% for sensitivity and specificity, respectively, using the SVM classifier.

In [35] the authors proposed a novel application based on deep convolution neural networks for automated diagnosis of fundoscopic images. The deep learning neural network was explicitly trained for automated grading of AMD and then the results generated by this algorithm were compared with the results obtained from another deep learning algorithm which used transfer learning and universal features. Further, the results generated by the two proposed methods were compared with results obtained by a trained clinical grader. In their proposed approach a two class classification problem was considered, in which normal/early cases were distinguished from intermediate/advanced stage AMD cases. Using different partitioning of data, several experiments were applied on a dataset of 130 000 images obtained from 4613 patients. Their proposed method achieved accuracies ranging between 88.4% and 91.6%.

A convolutional neural networks approach has been used in [36] for the classification of normal and AMD cases using 3195 fundoscopic images collected from the Ophthalmic Research Center, Iran University of Medical Sciences, and STARE datasets. As deep learning methods require a large dataset, different methods such as random crop and horizontal reflection methods have been used to increase the number of images in the dataset to improve the performance of learning algorithm. A convolutional neural network for deep learning was used in their proposed automated classification system and was formed of convolutional, pooling, and fully connected layers, where each layer was used to perform a different task. In the pooling layer, max-pooling as non-linear function was used for partitioning of the input image into different non-overlapping frames and return the maximum of each subregion as the output. This pooling layer was essentially used to reduce the spatial size of the input image, the number of parameters, and the computation in the network progressively to resist over-fitting. A rectified linear units (ReLU) neuron layer has been applied as an activation function in their proposed method. In the last max-pooling layer, several convolutions and high level reasoning are performed in the neural network via a fully connected layer. The neurons in the fully connected layer of their proposed system had full connections to all activations of the previous layers. The authors trained CNN using 15 epochs, where forward computing and error propagation were repeated 15 times using all training images. This all training process took almost 300 s on the local dataset and 38 s for the STARE dataset. The proposed automated system for drusen detection achieved accuracies of 95% and 81% on the local and STARE datasets, respectively.

In [37] Burlina *et al* proposed a new concept of image synthesizing for the generation of high-resolution fundus images. In their paper they used a deep learning method to synthesize a large dataset of artificial retinal fundoscopic images. As for the application of deep learning methods, for the diagnosis of different retinal diseases a large dataset is required to train deep convolution neural networks. The authors proposed a generative deep learning method for synthesizing a large dataset containing artificial images of different stages of AMD retinal disorder. This dataset can serve as a proxy dataset and can be used by deep learning machines and retinal specialists. Their proposed generative adversarial network was trained on 133 821 color fundoscopic images collected from 4613 participants from the AREDS study for AMD and non-AMD images. The application of deep learning algorithms can play a vital role in the design of automated diagnosis systems that can perform screening of retinal disorders and have clinical relevance with results the same as those observed for human performance, however, the use of a deep learning method requires a large dataset.

4.3 Retinal component extraction from OCT images

For the diagnosis of AMD from OCT images, we first need to extract retinal layers. From these extracted retinal layers, RPE segmentation is performed. Then further, we can identify drusens in the shape of bumps on the RPE layer of the retina from OCT brightness-scan [20, 32]. Different techniques that have been proposed in the literature for retinal layer extraction and drusen detection from OCT brightness-scan images are discussed in the following section.

4.3.1 Retinal layer extraction

In OCT brightness-scans, the precision of anatomical structures is a basic feature for the analysis and diagnosis of different macular and ocular diseases [38, 39]. Most commercially available OCT systems are designed mainly for the purpose of segmentation and to measure total retinal viscosity. A few systems also measure retinal nerve fiber layer (RNFL) viscosity with varying rates of accuracy [40]. The diagnosis of ocular diseases such as AMD, diabetic retinopathy, and glaucoma is based on the clinical examination and subjective study of OCT images by well-trained ophthalmologists.

The automated analysis of OCT images is an active field of research with the intention to accelerate diagnosis and empower remote identification of diseases [41]. Although advances are being made in the progress of automated retinal layer segmentation applications, for real-world cases clinical OCT images of severely damaged eyes with substantial pathology are divergent to the experimental images taken through long imaging sessions. These applications are still limited to a few restrictions [42] and also to the detection of eye diseases at an early phase before the appearance of severe pathology [43]. Fernández *et al* [44] proposed a model based upgrade segmentation approach with a blend of complex diffusion filtering with consistent enhancement diffusion filtering for de-noising, segmentation, and structural investigation in OCT brightness-scan retinal images. Their proposed method achieved good results in removing noise and the retinal layer extraction from OCT brightness-scan images was also improved.

In [47] Chiu *et al* proposed an autonomous system for retinal layer segmentation from OCT brightness-scan images. In their system they proposed the concept of the graph-cut hypothesis with a combination of dynamic-programming approaches that takes significantly less processing time for the segmentation of retinal layers and feature extraction. An enhanced adaptation of [47] dependent on a kernel-regression (KR) strategy was proposed in [48] to enhance the segmentation of retinal layer boundaries more accurately with a combination of the previously developed graph-theory and dynamic-programming (GTDP) framework. Algorithms were developed and tested with KR and KR-GTDP. The results showed that the KR based classification method did not perform well enough for accurate standalone retinal layer segmentation. The need to use the GTDP framework is the limitation of the proposed method. In [49] Duan *et al* proposed a new method for retinal layer segmentation from 2D and 3D OCT images based on the weighted geodesic distance derived from an eikonal equation using fast sweeping. The authors performed various experiments to compare their proposed method with manual segmentation of the retinal layers. Their proposed method segmented out retinal layers robustly from a complex retinal structure with irregularities and large curvature variations. The proposed method based on the distance from geodesic outperformed previously proposed methods with graph-based approaches and an active contour algorithm proposed for retinal layer segmentation from healthy and pathological images. In [50] a shortest path based graph search algorithm was proposed for automated segmentation of intra-retinal layers from OCT images. In their proposed method, based on a gradient, a minimal weight path between two graph nodes was estimated. The transition between pixel intensities was used to compute boundary position indices. The mean difference calculated between two boundary position indices of consecutive layers were used to quantify layer thickness. This procedure was repeated to segment out seven retinal layer boundaries with a mean computation time of 0.93 s.

4.3.2 Drusen detection from extracted retinal layers

After RPE layer extraction, the next step is the analysis of the RPE layer for the detection and segmentation of drusens. Qiang *et al* [51] proposed an algorithm for drusen extraction from OCT brightness-scan retinal images, which was influenced by the earlier learning of ordinary retinal morphology and structural features. A comparative analysis was presented by Nils *et al* in [16]. In their study, OCT images were compared with two other imaging technologies (i.e. FP and FA) regarding their sensitivity and specificity for AMD detection, CNV, and CNV activity. After analysing FA, FP, and SD-OCT volume scans of 120 eyes, the authors concluded that SD-OCT has high sensitivity for the detection of retinal diseases. However, this imaging technique cannot yet completely substitute the details which can be observed from FP and FA. As SD-OCT imaging technology is in the early stages of its use for the detection of retinal disease, this technology needs further advancements for the management of patients with retinal diseases. The authors state that SD-OCT was found to be beneficial in identifying and distinguishing drusens at an early stage.

Therefore, the SD-OCT imaging technology can definitely improve the quality of image understanding of eyes with AMD compared to fundoscopic images only. An image segmentation method has been proposed in [52] for RPE layer extraction and drusen shape detection from OCT brightness-scan images. The RNFL was also detected using the RPE layer. The RPE layer was also used to detect a blood bubble area in the RNFL complex. Finally, this model was used for the classification of the characteristics between two retinal disorders, DME and AMD. Sixteen SD-OCT brightness-scans were used for this classification. A filtering algorithm was used to reduce the noise present in the OCT brightness-scan images. RPE segmentation was performed after removing high value pixels from the RNFL complex. Drusen detection was performed on the RPE layer by calculating the top and bottom white pixel's mean point value for each column of the SD-OCT brightness-scan image and an estimated line was then drawn on the basis of the mean value. In the next step, segmentation of the RNF layer complex was performed. A threshold was used to find the RNFL complex, but both the RNFL and RPE layer were present in the result, therefore RPE layer segmentation was performed to remove the RPE layer from the resulting image. After performing this step, the RNFL complex and outer segment/inner segment (IS/OS) layers were presented in the result. Image segmentation was performed later on to detect bubbles in RNFL complex. In the next step, bubble classification was performed, as a bubble that appears in the IS/OS layers has the potential to be AMD not DME. Feature extraction was performed to further feed into a binary classifier to classify AMD and DME. This algorithm was then applied on unseen images where 149 bubbles of DME were found out of 172 in six DME images and 16 drusens were found out of 16 drusens in ten related macular degeneration images. This algorithm performed well only on black and white images.

Farsiu *et al* [61] characterized quantitative indicators (QI) to recognize drusens, indicating the stage of intermediate AMD from OCT brightness-scans. The QIs that were considered by the authors incorporate total retina (TR) volumes, RPE Drusen Complex (RPEDC) volumes, and irregular RPEDC stiffening volumes. They utilized a dataset of 384 patients, containing 115 images of elderly patients without AMD and 269 images of patients with intermediate AMD, representing the largest quantitative OCT images dataset to date. This OCT dataset contains RPE retinal layer stiffness profiles and other abnormalities. An irregular RPEDC stiffness QI was found to be the absolute most valuable single feature that can be used to diagnose the intermediate stage of AMD. A histogram-based approach in [62] was proposed for drusen detection. In the first stage, image enhancement was performed using a multi-level equalization method. Different color bands were utilized to analyse the image during the image enhancement process, since drusens can be extracted from the background based on morphology, structure, yellowish color, and brightness characteristics. It was observed that green band contains of more information about bright lesions than the other two bands (red and blue). The process of normalization was performed to remove irregular objects from the candidate region. For the segmentation of drusens from the background, a HALT operator was used. The experimental results of 23 samples show that the specificity and sensitivity for all samples was more than 96%, except one which had 86% because of the presence of noise.

In [63] Dufour *et al* proposed a method based on the random forest technique to locate drusens in the retinal pigment epithelium layer. In their proposed method, the authors used Bruch's membrane retinal layer segmentation by sampling a subsequent place around the retinal pigment epithelium layer by computing discriminative features for drusen classification. An enhanced version of the work presented in [63] is proposed in [64] by Olivira *et al.* The authors proposed multi-label classification based on new features using distance and a respective wavelet coefficient between the limiting boundaries for drusen location detection. Their proposed method detects individual drusens for further analysis of drusen size and area to categorize different stages of AMD. Their proposed method achieved an area under the ROC curve of 92.81% compared to the 92.03% achieved by Dufour's proposed method. An automated grading system for AMD was proposed in [65] based on the machine learning method bag of words (BoW). Based on this approach, the image contents were represented as a frequency vector histogram. The authors claim that their approach, based on a new machine learning method for the detection of different stages of AMD, performed well without the need for retinal layer segmentation and can help ophthalmologists to mass-screen AMD patients for the categorization of different AMD stages and proceed for further treatment. 130 000 fundoscopic images of the AREDS dataset were classified. Their proposed system achieved 88.4% accuracy.

4.3.3 AMD lesion detection

For the automated diagnosis of AMD from OCT images, most of the works have been done on retinal layer extraction and segmentation. Only a few methods have been proposed for the automated diagnosis of the intermediate stage of AMD. In [66] 3D OCT images were used for the identification of AMD. This method used a combination of oriented gradient and local binary pattern histograms to generate an effective volumetric representation based on a decomposition method. Pre-processing was performed to remove speckle noise by applying a split Bergman isotropic total variation algorithm in combination with a least squares approach. To improve the quality of the OCT images for volume of interest (VOI) extraction, and because of the curved nature of the retina, the OCT images were flattened using a second order polynomial least squares curve fitting. For feature extraction, the LBP histogram method was applied using a pre-processed volume for identification of a feature vector based on the local histogram. Further, identified feature vector sets were used for Bayesian network classifier training. This method was tested on 140 3D OCT images and six parameters, i.e. precision, *f*-score, sensitivity, specificity, accuracy, and area under curve (AUC), were calculated to evaluate its performance. Their proposed method achieved good accuracy when compared to previously proposed methods. In all cases, the Bayesian classifier was used to classify suspect images from normal ones. Their study only focused on three retinal layer segmentations of diseased eyes.

Severity analysis of AMD can be determined by quantifying drusen size. Therefore, in quantitative studies, retinal layer segmentation and drusen volume measurement are important, although the labor intensive segmentation of retinal layers consumes more time. A cross-sectional SD-OCT image view helps to characterize the retina, RPE layer, and also drusen complex in AMD affecting the eye. An automated algorithm was proposed in [73] that can be used for the efficient segmentation of the three retinal boundaries compared to manual segmentation. Guidelines have also been presented to identify the different layers of the retina that can be symptomatic of AMD progression, including RPE drusen complex and total retina. This algorithm was tested on 220 OCT brightness-scans of 20 patients. The average time of segmentation per image was noted as 1.7 s, compared to 3.5 min for manual segmentation. Their algorithm may not work properly when it is applied on images with disease progression from the intermediate stage of AMD to the advanced stage with a different rate of change in retinal pigment epithelium drusen complex. Their algorithm could also further segment out different types of drusens called soft drusen and sub-retinal drusen deposits that can be significant for the detection of AMD progression.

In [67] a technique for image classification based on deep learning of a neural network was proposed to differentiate AMD cases from normal cases using OCT brightness-scan images. Their study demonstrates that the use of OCT imaging technology is becoming the retinal imaging technology most commonly used by ophthalmologists for AMD retinal disease analysis with 4.39, 4.93, and 5.35 million OCT scans performed in 2012, 2013, and 2014, respectively, in the USA Medicare population. They used drusen presence, RPE atrophic changes, and intra-retinal and sub-retinal fluid as parameters to distinguish AMD cases from normal OCT images. For the training and classification system, a modified version of the VGG-16 convolution neural network was used as a deep learning model. An Electronic Medical Records (EMR) OCT database of 80 839 total OCT brightness-scans including 41 074 AMD and 39 765 normal scans were used to train the neural network and 20 163 scans including 11 616 AMD and 8547 normal scans were then used to validate the trained classifier. A neural network deep learning model was trained using 8000 iterations and then stopped because of an over-fitting problem. This deep learning module used an average time of 4.97 ms to evaluate a single OCT brightness-scan image. An ROC curve test evaluation was performed at the image level, macular level, and patient level. Their method achieved specificity and sensitivity of 91.54% and 84.63% respectively, with an accuracy of 87.63%. Peak sensitivity and specificity after constructing an ROC curve with optimal cutoffs were 87.08% and 87.05%, respectively. An area under the ROC curve of 92.77% was achieved by their classification system. The accuracy of their system could be improved if intra-retinal layer segmentation was also considered.

In [68] Zhao *et al* proposed an automated drusen detection system for early screening of AMD based on a multiple depth en-force method using OCT C-scans. Their proposed method detected drusen areas for all types of drusens from OCT scans without considering the need for segmentation of the RPE retinal layer. Their proposed method achieved good accuracy in detecting drusens on a dataset collected from the Casey Eye Institute of Oregon Health and Science University (OHSU). A method based on sparse coding and dictionary learning was proposed in [69] for the automated detection of AMD and RE from OCT images. The main focus of this study was to improve the state-of-the-art methods' classification performance. After pre-processing to crop and align retinal regions, using spatial pyramids and sparse coding, their approach obtained a global representation of the SD-OCT image. In a later stage classification was performed using an SVM. They tested their proposed approach on the Duke dataset and a locally collected dataset from Beijing. The proposed approach achieved an accuracy of 97% to classify SD-OCT scans from the Duke dataset. Using the locally collected dataset, their approach achieved an accuracy of 99%.

In [70] Arabi *et al* proposed a very simple method of drusen detection from the retinal pigment epithelium layer using OCT images. The authors used a simple image processing technique to extract the retinal pigment epithelium layer from OCT images. After layer extraction, sampling was performed. From each sample, the number of white pixels was counted and an averaged value was calculated for both types of normal and AMD affected images. From the calculated averaged value a threshold value was fixed based on a decision rule and images were classified.

Their proposed method achieved 75% accuracy. Different types of drusens are specific to different diseases, so as a future study there is a need for an algorithm that can classify drusens that are specific to different retinal disorders by utilizing different drusen segmentation techniques that have been proposed in the literature. A limitation of the proposed algorithms was their validity for only intermediate AMD detection and classification without considering proper retinal layer segmentation. In [71] Ji *et al* proposed an efficient deep learning method for automated diagnosis of ophthalmological pathology AMD and DME using SD-OCT images. The main intention of their proposed algorithm was to improve the classification by abridging the learning time. De-noising was performed using a block matching and 3D filtering approach. Then different morphological operations and linear fitting were performed for retinal layer (retinal subjects) detection. A modified version of the Inception V3 pre-trained model was used as a feature extractor. The extracted features were further used to feed as the input to the convolutional neural network for learning. The proposed method was tested on the Duke and locally collected datasets. For the comparison with previously proposed methods for retinal diseases detection, in this paper three experimental set-ups were adopted. In the first experimental set-up using the Duke dataset, a leave three out cross validation approach was used. The proposed method outperforms previously proposed methods. In the second experimental set-up using the Duke dataset, a comparison with a transfer based learning method was performed by dividing the dataset into 15 folds. Each fold contained scans from all three classes. The results generated by the proposed method were then compared with results produced by the previously proposed HOG based SVM and deep CNN classifiers methods. The proposed efficient deep learning method achieved an accuracy of 93%. In the third experimental set-up using the locally collected dataset, the authors performed classification based on OCT brightness-scan level. The proposed method achieved accuracy, sensitivity, and specificity of more than 95%.

4.4 Comparative analysis of diagnosis systems

Most of the automated systems proposed in the literature have focused on the segmentation of drusens and then the classification of AMD and non-AMD cases. Drusen segmentation techniques ignore those drusens which are very small in size and also consider other abnormalities as drusens, e.g. exudates and cotton wool spots. These abnormalities are due to other macular disorders such as central serous chorioretinopathy (CSCR) and retinal edema (RE) and their presence can cause false positive drusen detection. Therefore we need a technique that considers the whole macular region for the detection of drusenoid patterns while considering the discriminating characteristics of drusens used to distinguish these drusenoid patterns from exudates and cotton wool spots, leading to reliable diagnosis of AMD. For fundoscopic image analysis, the proposed automated diagnosis systems used different features for drusen segmentation to distinguish AMD cases from normal cases. Different features based on color, shape, and edge properties have been used in the literature [3, 4, 6–9, 31, 37] separately for automated diagnosis of AMD.

Researchers have performed drusen detection by considering the whole fundoscopic retinal image, whereas only drusen presence in the macular region can cause AMD. A summary of previously proposed automated diagnostic methods, using different image processing techniques, features sets, datasets, and performance details are presented in table 4.1 for fundoscopic image analysis. As one of the contributions of this chapter, we have proposed a system based on a combination of different features extracted only from the macular region for more accurate detection of AMD using fundoscopic images. In the recent era, the research paradigm has moved to the application of deep learning methods for automated screening of retinal abnormalities using fundoscopic retinal images [35, 36]. One limitation of the application of deep learning methods is the requirement for a large dataset. Different approaches have recently been proposed to compensate for this limitation of a lack of large datasets by applying deep generative models for high-resolution synthetic retinal image generation for automated analysis of AMD disease [37].

Table 4.2 gives an overview of the image processing methods that have been used for the purpose of retinal layers segmentation from normal and ophthalmic diseased OCT brightness-scan images. Most of these techniques have been used for retinal layer segmentation from normal images [47, 50, 53–55]. Some of the proposed techniques have been further used for RPE, RNFL, and inner limiting membrane layer segmentation from diseased images such as glaucoma, AMD, and DME [48, 49, 52, 56, 57] . From the literature review, it has been found that the graph-based method is the fastest and most accurate method of retinal layer segmentation from normal images. However, to perform retinal layer segmentation in the presence of lesions, this method needs further adjustments [47, 48]. In the literature various methods have been proposed for retinal layer segmentation from OCT images to date, but not a single method has been proposed which segments out all retinal layers perfectly from healthy as well as from pathological images. All the proposed techniques have their own failings, which provide research gaps for other researchers to work on layer segmentation methods that minimize intricacy and segmentation time, and facilitate easy implementation.

A summary of previously proposed automated diagnostic methods, using different image processing techniques from OCT images, and performance details are presented in table 4.3. Only a few methods have been proposed previously for the automated diagnosis of an intermediate stage of AMD without proper retinal layer segmentation and by using HOG and LBP features [32, 43], however, for reliable detection of drusens from the RPE layer, the proper retinal layer segmentation from OCT images is very important. We have proposed a new method based on the structure tensor for more accurate segmentation and extraction of all retinal layers. A comparative analysis of all the techniques discussed in this chapter has been published in [72].

4.5 Research limitations and challenges

From the literature review exploration of previously proposed systems for the automated diagnosis of AMD from fundoscopic and SD-OCT images, the following research gaps have been recognised:

Table 4.1. Techniques used for automated detection of AMD from fundoscopic images.

Reference	Method	Feature type	Feature set	Dataset (fundoscopic images)	Performance (Spec[1a]/Sens[2b]/Acc[3c]) [%]
Niemeijer *et al* [37]	Machine learning algorithm (KNN)	Statistical/color	Size, grey scale, green channel	300	88/77
Barriga *et al* [38]	Amplitude modulation, frequency modulation, partial least squares, principal component analysis	—	Retinal background, vessels, hard and soft drusen	Wisconsin Fundoscopic Photo Reading Centre's website	100
Friberg *et al* [45]	Histogram-based adaptive local thresholding	Statistical	Size, area, number of drusens	349 (IRIDEX Corporation, Mountain View, CA)	—
Hijazi *et al* [9]	Histogram-based approach (dynamic time wrapping)	Color/intensity	RGB channels and HSI	144	62/83/69
Jayanthi *et al* [6]	Auto-associative neural network classifier	Statistical/structural	Texture-based shape analysis, namely statistical and structural	100	—
Priya *et al* [7]	Probabilistic neural network (PNN), discrete wavelet transform, Kirsch edge detection algorithm	Statistical	Area, mean, standard deviation, radius, perimeter, variance, the entropy of the pre-processed image	300	94/95
Soliz *et al* [8]	Independent component analysis	Color/intensity	Pixel intensities and color	2500	—
Zheng *et al* [31]	Image de-noising (with non-linear means filter, morphological operations), Ada-LS-SVM classifier (pixel wise classification, region-wise classification)	Color/intensity	Pixel-wise classification (hue histogram and color histogram), region-wise classification	CAPT study 50 images, and Amish study 88 images	For CAPT CFP82/75/82, for Amish CFP86/71/85
Kumary *et al* [46]	Drusen detection based on illumination correction	Color/intensity	Pixel-wise analysis for drusen detection and region based drusen area quantification	STARE and ARIA dataset	—
Acharya *et al* [34]	Automated screening of dry and wet AMD	Statistical/structural	PHOG	Local dataset containing 945 images	87.2/80/85.1
Burlina *et al* [35]	Deep convolutional neural networks (DCNN)	—	—	130 000 (AREDS) dataset	Accuracy ranged between 88.4 to 91.6
Langarizadeh *et al* [36]	Decision support system for AMD using CNN	—	CNN	Local and STARE	95/81
Burlina *et al* [37]	Deep generative models for a high-resolution synthetic retinal images dataset for AMD	—	—	133 821 (AREDS) dataset	—

[a] Specificity
[b] Sensitivity
[c] Accuracy

Table 4.2. Image processing techniques used for retinal layer segmentation.

Reference	Layer segmentation method	Dataset (OCT images)
Bagci *et al* [53]	Grey level lapping, edge detector operator	10 normal images
Rossant *et al* [54]	Diffusion filter, Markov random field, *K*-mean clustering	25 normal, 47 diseased images
Yang *et al* [56]	Canny edge, dynamic programming	19 glaucomatic images of the left and right eyes
Yazdanpanah *et al* [55]	Chan–Vese energy-minimizing active contour	—
Chiu *et al* [47]	Graph theory and dynamic programming	Normal images
Wilkins *et al* [58]	Median filter, bilateral filter, thresholding and boundary detection algorithm	16 patients affected by macular edema
Huang *et al* [57]	Canny edge operator	4 normal and 24 diabetic macular edema (DME) images
Kafieh *et al* [59]	Graph-based segmentation	Normal images
Abhishek *et al* [52]	Median filter, method I: canny edge detection filter, LOG filter, method II: graph theory	DME images
Chiu *et al* [48]	Kernel regression and GTDP (graph theory + dynamic programming)	110 brightness-scan DME images
Stankiewicz *et al* [60]	Graph theory	15 images
Duan *et al* [49]	Weighted geodesic distance derived from an eikonal equation via fast sweeping	30 healthy, 20 AMD
Roy *et al* [50]	Boundary position indices based on graph theory method	25 normal images from the Medical Research Foundation, a unit of Sankara Nethralaya in Chennai, India

1. Most of the work in fundus imaging was done to detect drusens based on segmentation. Descriptor based methods mostly considered the whole image for analysis. Drusen presence in various places on the retina can be caused by different retinal diseases. Drusen presence in the macular region only can cause AMD. Therefore whole image analysis can cause false drusen detection.
2. There is no dataset available containing early and advanced AMD cases which leads to a gap for the analysis of fundoscopic images to diagnose AMD at an early stage.

Table 4.3. Techniques used for the automated detection of AMD from OCT images.

Reference	Purpose	Method	Features	Dataset (SD-OCT images)	Performance
Albarrak *et al* [32]	Classification of intermediate AMD	Feature based detection	HOG, LBP	140	—
Srinivasan *et al* [43]	Classification of intermediate AMD	Feature based detection	HOG	45	95% accuracy
Farsiu *et al* [61]	Quantitative and classification of intermediate AMD	Graph theory to extract retinal layer boundaries	Thickness details of RPE, RPEDC, and TR boundaries	38 400	99% precision from controlled subjects
Sugruk *et al* [62]	RPE layer segmentation for detection of drusen shape	Drusen segmentation: mean point values of the top and the bottom white pixel of the RPE layer	—	16	—
Lee *et al* [67]	Classification of AMD	VGG-16 CNN as deep learning model	—	52 690 normal and 48 312 AMD	92.64%/93.69% sensitivity/specificity
Venhuizen *et al* [65]	Grading of AMD	Bag of words (BoW)	Image represented as frequency vector (histogram) of visual word	3256	98.2%/91.2% sensitivity/specificity
Sun *et al* [69]	Automated macular pathology detection	Sparse coding and dictionary learning	—	Duke/local	97%/99% accuracy
Arabi *et al* [70]	Drusen detection from RPE layer	Decision rule based on averaged mean threshold value	Averaged mean value, thresholding	16	75% accuracy
Ji *et al* [71]	Retinal disease detection	Deep learning	—	Duke/local dataset	97.8%/98.9%/98.5% sensitivity/specificity/accuracy (averaged)

3. Most of the previously proposed work has been done for the segmentation of only a few retinal layers.
4. Most of the results generated by previously proposed retinal layer segmentation methods are available on the extraction of layers for normal scans.
5. No quantification of AMD cases according to clinical standards has been proposed in the literature.
6. There is no gold standard dataset available for the combined analysis of SD-OCT and fundoscopic images.

The key challenges found for automated diagnosis of AMD from fundoscopic and SD-OCT images are as follows:

1. AMD prediction at an early stage remains an open problem but may have a significant impact on the prevention of blindness due to AMD.
2. One of the main challenge is the consideration of drusen presence area in fundoscopic images.
3. Drusen discrimination from other macular abnormalities s a challenging problem and can cause false drusen detection in fundoscopic images.
4. Accurate extraction of the RPE retinal layer and its thickness and thinness details from SD-OCT images is a key challenge for the prediction of AMD from SD-OCT images.
5. Discrimination of early suspects and advanced AMD cases from SD-OCT images is challenging.
6. The grading of AMD into different stages using both the fundoscopic and SD-OCT imaging modalities is a challenging task from clinical perspective.

4.6 Conclusion

From the literature review, it can be seen that most AMD diagnostic methods have been proposed for fundoscopic images. Fundoscopic imaging is complicated by the facts that the resulting image set-up is technically challenging and has historically involved comparatively expensive equipment and a highly trained ophthalmic photographer. Many automated systems have been proposed to make fundoscopic imaging more accessible, resulting in less dependence on such experience and expertise. Most of the existing algorithms focus on segmentation of candidate drusen regions first and then classifying these regions as drusens or non-drusens. Most of the automated systems are developed for drusen detection and quantification from fundoscopic images for the diagnosis of the intermediate stage of AMD. In the diagnosis of AMD, the use of OCT imaging technology is also found to be helpful. OCT is so far the only imaging technique that has made accessible the cross-sectional details of the retina and choroid, where most of the AMD signs that appear in an early stage can be seen clearly. Therefore, for timely diagnosis and treatment of AMD the OCT imaging technique is found to be the most helpful and current research is in progress for developing an automated system for accurate and early diagnosis of AMD. For developing automated diagnostic systems, the main emphasis is on accurate detection and segmentation of different retinal layers, which can further help to measure the thickness and thinness of each layer. These measurement details can then be used further in the diagnosis of different retinal disorders.

References

[1] Williams G A, Scott I U, Haller J A, Maguire A M, Marcus D and McDonald H R 2004 Single-field fundus photography for diabetic retinopathy screening: a report by the American Academy of Ophthalmology *Ophthalmology* **111** 1055–62

[2] MPS Group 1991 Laser photocoagulation of subfoveal recurrent neovascular lesions in age-related macular degeneration: results of a randomized clinical trial *Archiv. Ophthalmol.* **109** 1232–41

[3] A-REDSR Group 2003 Potential public health impact of age-related eye disease study results: AREDS report no. 11 *Archiv. Ophthalmol.* **121** 1621

[4] Hageman G S, Luthert P J, Chong N V, Johnson L V, Anderson D H and Mullins R F 2001 An integrated hypothesis that considers drusen as biomarkers of immune-mediated processes at the RPE-Bruch's membrane interface in aging and age-related macular degeneration *Progr. Retinal Eye Res.* **20** 705–32

[5] Raza G, Rafique M, Tariq A and Akram M U 2013 Hybrid classifier based drusen detection in colored fundus images *IEEE Jordan Conf. on Applied Electrical Engineering And Computing Technologies* pp 1–5

[6] Jayanthi D, Devi N and SwarnaParvathi S 2010 Automatic diagnosis of retinal diseases from color retinal images, arXiv:1002.2408

[7] Priya R and Aruna P 2011 Automated diagnosis of age-related macular degeneration from color retinal fundus images *3rd Int. Conf. on Electronics Computer Technology* pp 227–30

[8] Soliz P, Russell S, Abramoff M, Murillo S, Pattichis M and Davis H 2008 Independent component analysis for vision-inspired classification of retinal images with age-related macular degeneration *IEEE Southwest Symp. on Image Analysis and Interpretation* pp 65–8

[9] Hijazi M H A, Coenen F and Zheng Y 2013 Age-related macular degeneration screening using data mining approaches *1st Int. Conf. on Artificial Intelligence, Modelling and Simulation* pp 299–303

[10] Burlina P, Freund D E, Dupas B and Bressler N 2011 Automatic screening of age-related macular degeneration and retinal abnormalities *Annual Int. Conf. of the IEEE Engineering in Medicine and Biology Society* pp 3962–6

[11] Chowdhury A R and Banerjee S 2012 Cellular Neural Network based algorithm in image analysis of age related macular degeneration *IET Chennai 3rd Int. on Sustainable Energy and Intelligent Systems (SEISCON)* 1–4

[12] Gonzalez R C and Wintz P 1977 *Digital Image Processing* (Reading, MA: Addison-Wesley) p 451

[13] González R C, Woods R E and Eddins S L 2004 *Digital Image Processing Using MARLAB* (London: Pearson)

[14] Chang S G, Yu B and Vetterli M 2000 Spatially adaptive wavelet thresholding with context modeling for image denoising *IEEE Trans. Image Process.* **9** 1522–31

[15] Wilson M P, Yang S, Mitra S, Raman B, Nemeth S C and Soliz P 2003 Full automation of morphological segmentation of retinal images: a comparison with human-based analysis *Proc. SPIE Medical Imaging 2003: Image Processing* **5032** 27–38
[16] Akman A 2018 Optical coherence tomography: manufacturers and current systems *Optical Coherence Tomography in Glaucoma* (Berlin: Springer) pp 27–37
[17] Brandon L and Hoover A 2003 Drusen detection in a retinal image using multi-level analysis *Int. Conf. on Medical Image Computing and Computer-assisted Intervention* pp 618–25
[18] Bhuiyan A *et al* 2013 Drusen detection and quantification for early identification of age related macular degeneration using color fundus imaging *J. Clin. Exp. Ophthalmol* **4** 2
[19] Santos-Villalobos H *et al* 2011 Statistical characterization and segmentation of drusen in fundus images *Annual Int. Conf. of the IEEE Engineering in Medicine and Biology Society* pp 6236–41
[20] Waseem S, Akram M U and Ahmed B A 2014 Drusen exudate lesion discrimination in colour fundus images *14th Int. Conf. on Hybrid Intelligent Systems* pp 176–80
[21] Sbeh Z B, Cohen L D, Mimoun G, Coscas G and Soubrane G 1997 An adaptive contrast method for segmentation of drusen *Proc. Int. Conf. on Image Processing* pp 255–8
[22] Sbeh Z B, Cohen L D, Mimoun G and Coscas G 2001 A new approach of geodesic reconstruction for drusen segmentation in eye fundus images *IEEE Trans. Med. Imaging* **20** 1321–33
[23] Rapantzikos K and Zervakis M 2001 Nonlinear enhancement and segmentation algorithm for the detection of age-related macular degeneration (AMD) in human eye's retina *Proc. 2001 Int. Conf. on Image Processing* pp 1055–8
[24] Thdibaoui A, Rajn A and Bunel P 2000 A fuzzy logic approach to drusen detection in retinal angiographic images *Proc. 15th Int. Conf. on Pattern Recognition* pp 748–51
[25] Quellec G, Russell S R and Abràmoff M D 2011 Optimal filter framework for automated, instantaneous detection of lesions in retinal images *IEEE Trans. Med. Imaging* **30** 523–33
[26] Zhang W, Yamamoto K and Hori S 2008 Optical coherence tomography for assessment of diabetic macular edema *Int. J. Ophthalmol.* **1** 370–3
[27] Rani P and Kaleka J 2015 Macular degeneration detection in retinal fundus images using differential algorithm, *Int. J. Innov. Res. Sci., Eng. Technol.* **4** 7741–6
[28] Kim Y J and Kim K G 2018 Automated segmentation methods of drusen to diagnose age-related macular degeneration screening in retinal images *Comput. Math. Methods Med.* **2018** 6084798
[29] Mohaimin S M, Saha S K, Khan A M, Arif A S M and Kanagasingam Y 2018 Automated method for the detection and segmentation of drusen in colour fundus image for the diagnosis of age-related macular degeneration *IET Image Proc.* **12** 919–27
[30] Rubin D L and de Sisternes Garcia L 2017 Method and system for evaluating progression of age-related macular degeneration *US Patent* US20160174830
[31] Zheng Y *et al* 2013 An automated drusen detection system for classifying age-related macular degeneration with color fundus photographs *IEEE 10th Int. Symp. on Biomedical Imaging* pp 1448–51
[32] Albarrak A, Coenen F and Zheng Y 2013 Age-related macular degeneration identification in volumetric optical coherence tomography using decomposition and local feature extraction *Proc. 2013 Int. Conf. on Medical Image, Understanding and Analysis* pp 59–64
[33] Phan T V, Seoud L, Chakor H and Cheriet F 2016 Automatic screening and grading of age-related macular degeneration from texture analysis of fundus images *J. Ophthalmol.* **2016** 5893601

[34] Acharya U R *et al* 2017 Automated screening tool for dry and wet age-related macular degeneration (ARMD) using pyramid of histogram of oriented gradients (PHOG) and nonlinear features *J. Comput. Sci.* **20** 41–51

[35] Burlina P M, Joshi N, Pekala M, Pacheco K D, Freund D E and Bressler N M 2017 Automated grading of age-related macular degeneration from color fundus images using deep convolutional neural networks *JAMA Ophthalmol.* **135** 1170–6

[36] Langarizadeh M, Maghsoudi B and Nilforushan N 2017 Decision support system for age-related macular degeneration using convolutional neural networks *Iran. J. Med. Phys.* **14** 141–8

[37] Burlina P M, Joshi N, Pacheco K D, Liu T A and Bressler N M 2019 Assessment of deep generative models for high-resolution synthetic retinal image generation of age-related macular degeneration *JAMA Ophthalmol.* **137** 258–64

[38] Shen S Y *et al* 2008 The prevalence and types of glaucoma in Malay people: the Singapore Malay Eye Study *Investigative Ophthalmol. Vis. Sci.* **49** 3846–51

[39] Huang D *et al* 1991 Optical coherence tomography *Science* **254** 1178–81

[40] Fujimoto J G, Drexler W, Schuman J S and Hitzenberger C K 2009 Optical coherence tomography (OCT) in ophthalmology: introduction *Opt. Express* **17** 3978–9

[41] Han I C and Jaffe G J 2010 Evaluation of artifacts associated with macular spectral-domain optical coherence tomography *Ophthalmology* **117** 1177–89

[42] Lee N, Laine A F and Smith T R 2008 Learning non-homogenous textures and the unlearning problem with application to drusen detection in retinal images *5th IEEE Int. Symp. on Biomedical Imaging: From Nano to Macro* pp 1215–8

[43] Srinivasan P P *et al* 2014 Fully automated detection of diabetic macular edema and dry age-related macular degeneration from optical coherence tomography images *Biomed. Opt. Express* **5** 3568–77

[44] Fernández D C and Salinas H 2005 Extracting subretinal layers on stratus OCT images via a structure tensor approach combined with a nonlinear diffusion process *Investigative Ophthalmol. Vis. Sci.* **46** 2575

[45] Friberg T R, Huang L, Palaiou M and Bremer R 2007 Computerized detection and measurement of drusen in age-related macular degeneration *Ophthal. Surg., Lasers Imaging Retina* **38** 126–34

[46] Kumari K and Mittal D 2015 Automated drusen detection technique for age-related macular degeneration *J. Biomed. Eng. Med. Imaging* **2** 18

[47] Chiu S J, Toth C A, Rickman C B, Izatt J A and Farsiu S 2012 Automatic segmentation of closed-contour features in ophthalmic images using graph theory and dynamic programming *Biomed. Opt. Express* **3** 1127–40

[48] Chiu S J, Allingham M J, Mettu P S, Cousins S W, Izatt J A and Farsiu S 2015 Kernel regression based segmentation of optical coherence tomography images with diabetic macular edema *Biomed. Opt. Express* **6** 1172–94

[49] Duan J, Tench C, Gottlob I, Proudlock F and Bai L 2017 Automated segmentation of retinal layers from optical coherence tomography images using geodesic distance *Pattern Recognit.* **72** 158–75

[50] Roy P, Gholami P, Parthasarathy M K, Zelek J and Lakshminarayanan V 2018 Automated intraretinal layer segmentation of optical coherence tomography images using graph-theoretical methods *Proc. SPIE* **10483** 104832U

[51] Chen Q *et al* 2013 Automated drusen segmentation and quantification in SD-OCT images *Med. Image Anal.* **17** 1058–72

[52] Abhishek A M, Berendschot T T, Rao S V and Dabir S 2014 Segmentation and analysis of retinal layers (ILM & RPE) in optical coherence tomography images with edema *IEEE Conf. on Biomedical Engineering and Sciences* pp 204–9

[53] Bagci A M, Shahidi M, Ansari R, Blair M, Blair N P and Zelkha R 2008 Thickness profiles of retinal layers by optical coherence tomography image segmentation *Am. J. Ophthalmol.* **146** 679–87

[54] Rossant F, Ghorbel I, Bloch I, Paques M and Tick S 2009 Automated segmentation of retinal layers in OCT imaging and derived ophthalmic measures *IEEE Int. Symp. on Biomedical Imaging: From Nano to Macro* pp 1370–3

[55] Yazdanpanah A, Hamarneh G, Smith B R and Sarunic M V 2011 Segmentation of intra-retinal layers from optical coherence tomography images using an active contour approach *IEEE Trans. Med. Imaging* **30** 484–96

[56] Yang Q *et al* 2010 Automated layer segmentation of macular OCT images using dual-scale gradient information *Opt. Express* **18** 21293–307

[57] Huang Y *et al* 2013 Development of a semi-automatic segmentation method for retinal OCT images tested in patients with diabetic macular edema *PLoS One* **8** e82922

[58] Wilkins G R, Houghton O M and Oldenburg A L 2012 Automated segmentation of intraretinal cystoid fluid in optical coherence tomography *IEEE Trans. Biomed. Eng.* **59** 1109–14

[59] Kafieh R, Rabbani H and Kermani S 2013 A review of algorithms for segmentation of optical coherence tomography from retina *J. Med. Signals Sensors* **3** 45

[60] Stankiewicz A, Marciniak T, Dąbrowski A, Stopa M, Rakowicz P and Marciniak E 2016 Improving segmentation of 3D retina layers based on graph theory approach for low quality OCT images *Metrol. Meas. Syst.* **23** 269–80

[61] Farsiu S *et al* 2014 Quantitative classification of eyes with and without intermediate age-related macular degeneration using optical coherence tomography *Ophthalmology* **121** 162–72

[62] Sugmk J, Kiattisin S and Leelasantitham A 2014 Automated classification between age-related macular degeneration and diabetic macular edema in OCT image using image segmentation *7th Biomedical Engineering Int. Conf.* pp 1–4

[63] Dufour P A, De Zanet S, Wolf-Schnurrbusch U and Kowal J 2012 Classification of drusen positions in optical coherence tomography data from patients with age-related macular degeneration *Proc. 21st Int. Conf. on Pattern Recognition* pp 2067–70

[64] Oliveira J, Gonçalves L, Ferreira M and Silva C A 2017 Drusen detection in OCT images with AMD using random forests *IEEE 5th Portuguese Meeting on Bioengineering* pp 1–4

[65] Venhuizen F G *et al* 2017 Automated staging of age-related macular degeneration using optical coherence tomography *Invest. Ophthalmol. Vis. Sci.* **58** 2318–28

[66] Chiu S J, Izatt J A, O'Connell R V, Winter K P, Toth C A and Farsiu S 2012 Validated automatic segmentation of AMD pathology including drusen and geographic atrophy in SD-OCT images *Invest. Ophthalmol. Vis. Sci.* **53** 53–61

[67] Lee C, Baughman D and Lee A 2017 Deep learning is effective for classifying normal versus age-related macular degeneration optical coherence tomography images *Ophthalmol. Retina* **1** 322–7

[68] Zhao R *et al* 2017 Automated drusen detection in dry age-related macular degeneration by multiple-depth, en face optical coherence tomography *Biomed. Opt. Express* **8** 5049–64

[69] Sun Y, Li S and Sun Z 2017 Fully automated macular pathology detection in retina optical coherence tomography images using sparse coding and dictionary learning *J. Biomed. Opt.* **22** 016012

[70] Arabi P M, Krishna N, Ashwini V and Prathibha H 2018 Identification of age-related macular degeneration using OCT images *IOP Conf. Ser.: Mater. Sci. Eng.* **310** 012096
[71] Ji Q, He W, Huang J and Sun Y 2018 Efficient deep learning-based automated pathology identification in retinal optical coherence tomography images *Algorithms* **11** 88
[72] Khalid S, Khalil T and Nasreen S 2014 A survey of feature selection and feature extraction techniques in machine learning *Science and Information Conf.* pp 372–8
[73] Tariq A and Akram M 2010 An automated system for colored retinal image background and noise segmentation *IEEE Symp. on Industrial Electronics and Applications (ISIEA)* 423–7

IOP Publishing

Photo Acoustic and Optical Coherence Tomography Imaging, Volume 2
Fundus imaging for the retina
Ayman El-Baz and Jasjit S Suri

Chapter 5

Interobserver variability in the determination of diabetic retinopathy and quality of fundus image

Sílvia Rêgo, Matilde Monteiro-Soares, Filipe Soares and Marco Dutra-Medeiros

5.1 Introduction

Diabetic retinopathy affects the retina, the neurosensory tissue in the eye that plays an essential role in human vision [1, 2]. The retina contains photoreceptors, the cells that sense light and convert optical images into electrical signals that our brain can understand [1, 2]. Many small blood vessels nourish the photoreceptors in the retina [1, 2]. These blood vessels become damaged over time by high blood sugar levels [3]. Such lesions are modifications in the retinal vessels and encompass microaneurysms, intraretinal microvascular anomalies, venous beading, haemorrhages, and neovascularization [3]. Identification of these early and asymptomatic lesions and subsequent treatment of the condition is essential to avoid blindness in patients with diabetes [3–5]. Retinopathy is the most common diabetes-related microvascular complication [6–8], occurring in up to 35% of individuals [9]. There is evidence that identifying and treating people with early diabetic retinopathy leads to better outcomes than treating the disease at later stages, when people already present visual loss [8].

The gold standard test to screen for diabetic retinopathy is a visual analysis of eye fundus photographs for identification of microvascular lesions [3], which should be described following the International Clinical Disease Severity Scale (ICDSS) [10]. This classification is based on the presence and severity of lesions to differentiate proliferative from nonproliferative diabetic retinopathy, and categorise each one as mild, moderate, or severe [10]. Prior studies have shown that the classification of fundus images by ophthalmologists using this scale is insufficiently reliable, as indicated by kappa coefficients that ranged from 0.49 to 0.67 [11, 12].

Furthermore, it is reported in the literature that image-based diagnosis is subject to interobserver variability, which in part depends on the quality of images

doi:10.1088/978-0-7503-2056-6ch5

presented on the display [13]. Therefore, difficulties in the visual perception and interpretation of fundus images influence the effectiveness of diabetic retinopathy telescreening [13] and could result in misdiagnosis and visual loss for a patient. Moreover, there are few studies about the effect of image quality on diabetic retinopathy image grading.

5.2 Fundus imaging

Fundus imaging consists of taking photographs of the fundus of the eye through the pupil [2, 14]. A fundus camera, comprising a specialised camera attached to a low-power microscope, enables the taking of fundus photographs in a non-invasive manner [2, 14]. The fundus camera's illuminating and imaging light rays enter and exit the pupils, creating fundus photographs [13]. A fundus photograph is a 2D representation of the 3D semitransparent retinal tissues and shows the eye structures: the optic disc, the macula, and the blood vessels [2].

5.2.1 The diabetic retinopathy classification process

Currently, telescreening is the most commonly used screening method for diabetic retinopathy; therefore, fundus images are essential for ophthalmologists to identify this clinical condition [5]. Many countries, including Portugal and the UK, have implemented telemedicine-based diabetic retinopathy screening [5]. This method enables the systematic screening of patients with diabetes while dealing with a shortage of ophthalmologists [5]. All individuals with diabetes must undergo this screening annually—three to five years after type 1 diabetes diagnosis, and once type 2 diabetes is diagnosed [15].

Digital fundus images are acquired at primary care units or in the community by trained technicians [5]. Ophthalmologists receive all the images at reading centre, located at a hospital, and visually inspect them to identify the clinical findings described in the Early Treatment Diabetic Retinopathy Study (ETDRS) [5]. In most screening programmes, each digital fundus image is graded according to the ICDSS [10]. Microaneurysms, intraretinal haemorrhages, and focal areas of retinal ischemia, also known as cotton-wool spots, are early signs of diabetic retinopathy, a stage classified as nonproliferative diabetic retinopathy (NPDR) [16]. Depending on the type and number of these pathological structures, ophthalmologists classify NPDR as mild or moderate [16]. If the image shows venous beading, intraretinal microvascular abnormalities, and more severe haemorrhages, diabetic retinopathy (DR) has progressed to the stage of severe NPDR [16]. The presence of neovessels, i.e., fragile new blood vessels at the inner surface of the retina, indicates that retinopathy has progressed to proliferative [16]. This stage usually results in vision loss, as the new blood vessels often bleed, causing vitreous haemorrhage [16]. Notably, hard yellow exudates can be present in fundus images at any stage of the disease [16]. This stage is known as diabetic macular oedema and may cause vision loss [16]. The classification will allow each image to be assigned to one of the categories: no DR, mild NPDR, moderate NPDR, severe NPDR, PDR, or diabetic macular oedema.

Screened individuals whose images are graded as more than mild NPDR or diabetic macular oedema require further assessment and are therefore referred to ophthalmology [2]. Image classification is a time-consuming task for ophthalmologists, as all images, usually two images per eye of each patient [5], must be analysed and graded.

As explained above, the diagnosis of diabetic retinopathy is an image-based process and involves the tasks of detecting and estimating the volume of an object in an image. The most common objects of interest for this condition are retinal vessels, the optic disc, optic disc centre, macula, fovea, haemorrhages, microaneurysms, cotton-wool spots, and hard exudates [17]. Therefore, ophthalmologists must receive images in which objects are sufficiently visible.

5.2.2 Fundus image quality

The function of fundus images is to display the fundus anatomy in a manner that the information contained in that representation enables the reader, i.e., the ophthalmologist, to perform the classification and interpretation and provide an adequate screening result [18]. Fundus images that, due to poor quality, make the reading for a screening result unfeasible are classified as not classifiable, meaning the patient needs to undergo assessment in an ophthalmology appointment. This is the case for images obscured by cataracts, for example. On the other hand, if image quality prevents images from clearly exhibiting these structures or even not displaying them at all, classification becomes more challenging, and lesions may pass unnoticed.

The ability of an observer to detect signs of disease in images depends on three major factors: image quality, viewing conditions, and observer performance [18]. The patient's ocular and vision characteristics, as well as the patient's collaboration regarding body immobilisation during the acquisition of images, also influence the quality of fundus images [18]. Equally, the characteristics of the imaging equipment, the imaging procedure and the skill of the operator influence the image quality [18]. The parameters of image quality related to the acquisition procedure are already known [18]. Therefore, ophthalmic photographers use standardised imaging protocols to ensure good image quality [19]. Despite its importance, less effort has been made to understand fundus image quality as perceived by ophthalmologists. Indeed, there are no established or commonly used criteria in clinical practice for ophthalmologists to judge the quality of fundus images.

The ability of observers to see objects in medical images depends on viewing conditions such as contrast, background brightness, object size, viewing distance, glare, and background structure [18]. For this reason, imaging systems, namely display hardware and software, have evolved in recent years [18].

Factors involved in observer performance, such as expertise and education, as well as ambient-related factors (e.g., interruptions and sequential context bias), have been extensively studied [18]. However, the evaluation of tissues in medical images is also affected by subjective interpretation, which influences observer performance.

5.3 Interobserver variability and reliability of DR classification

Subjective interpretation of imaging findings translates into interobserver variability [20]. Variability between ophthalmologists in diabetic retinopathy screening relates to screening error and may compromise screening results. Ultimately, interobserver variability may compromise the patient's vision outcomes. Few studies have assessed interobserver variability in DR screening, however. The degree of interobserver variation is quantified by the extent to which observers agree with each other about disease findings and is represented mathematically by percent agreement [21, 22]. Therefore, studies assessing interobserver variability have the same medical images classified by two or more observers [21, 22]. The agreement is based on each observer assigning each image to one of two or more non-overlapping categories [21, 22]. The resulting data is reported in a contingency table, in which the sum of the probabilities in the four diagonal cells represents the proportion of agreement [21, 22]. To better understand the magnitude of proportion agreement, it is useful to consider the reverse, i.e., the disagreement among the observers [23].

Reliability quantifies the ratio of variability between scores of the same patients to the total variability of all scores in the sample [21, 22]. In diagnostic studies, reliability expresses the ability of the scores to distinguish between individuals with the presence or absence of a disease or condition [21, 22]. Kappa is the appropriate statistic if the outcome is nominal data; weighted kappa, if ordinal data [23]. When assessing the reliability between two observers, Cohen's kappa is indicated; between three or more observers, Fleiss' kappa should be used [23]. When assessing the reliability for all observer pairs in a study with three or more observers, then Light's kappa is the statistic measure to calculate [24]. Kappa values can range from −1 to +1 [23]. A kappa value of one indicates perfect agreement between the raters; a kappa result of zero indicates agreement due only to pure chance [23]. Negative kappa values indicate worse disagreement than expected by chance [23]. There is not a consensus about the interpretation of kappa values between 0 and 1. According to Cohen, the interpretation of kappa results should be made as follows: $\leqslant 0$ no agreement; 0.01 to 0.20, none to slight agreement; 0.21 to 0.40, fair; 0.41 to 0.60, moderate; 0.61 to 0.80, substantial; and 0.81 to 1.00, almost perfect [23]. McHugh have proposed the following for interpreting the magnitude of kappa: 0 to 0.20, no agreement; 0.21 to 0.39, minimal; 0.40 to 0.59, weak; 0.60 to 0.79, moderate; 0.80 to 0.90, strong; and above 0.90, almost perfect [23]. Thus, McHugh suggested that a kappa value below 0.60 indicates inadequate agreement among the observers [23]. These interpretations are somewhat arbitrary, but it is generally accepted that the closer the kappa value is to 1, the greater the reliability; no other interpretation will be given regarding the evidence of the reliability of diabetic retinopathy classification that follows. Kappa values are affected by both the prevalence [22] and heterogeneity of the study sample [25]. On the other hand, the percent agreement is affected by the agreement by chance [21, 22].

Variability in the classification of DR between ophthalmologists has been examined in a few previous studies [11, 12, 26]. All studies used kappa, but only one study reported overall percent agreement. In these studies, it is common to evaluate multiprofessional groups, including ophthalmologists, nurses, general practitioners, and ophthalmic photographers, as well as diverse professional groups. For this purpose, we only considered studies that assessed ophthalmologists or diverse professional groups that included ophthalmologists.

There is evidence of insufficient reliability for the ETDRS classification scale by ophthalmologists of single-field digital eye fundus images; three studies reported kappa values around 0.60 for the presence of diabetic retinopathy, as well as diabetic macular oedema, or to decide about the need for referral [11, 12, 26]. The study by Gegundez *et al* focused on the variability between three ophthalmologists; the study by Ruamviboonsook *et al* an analysis of the classification by three retina specialists; and our research team's previous study included two ophthalmologists (one retina specialist and one general ophthalmologist) [11, 12, 26]. These studies provide insights into the extent to which ophthalmologists differentiate diabetic retinopathy findings. A direct comparison of the results of reliability studies for the classification of diabetic retinopathy is difficult, however, as different professional categories of pairs of readers were used between studies and different prevalences were reported. Also, there are different modifications to the ETDRS scale.

In a previous study by our research team, two ophthalmologists classified and interpreted fundus images for diabetic retinopathy [26]. Interrater agreement for diabetic retinopathy, maculopathy, and referral cases were 85%, 96%, and 94%, respectively [26]. For the absence of diabetic retinopathy, referable diabetic retinopathy and maculopathy, the interrater agreement was considerably higher than for its presence (proportion of agreement of 91% versus 56%, 98% versus 56%, and 97% versus 57%, respectively) [26]. These results suggest greater difficulty in detecting diabetic retinopathy lesions than in deciding that an image is clean, mainly due to the small size of lesions and difficulty in differentiating them from other findings.

5.4 Image quality

In diabetic retinopathy telescreening, the quality of the retinal images is essential to adequately identify disease. However, as far as we know, only one study has dedicated itself to the understanding of the elements involved in low reliability [26]. In addition, it is recognised that a diagnostic test only provides useful information if it yields reliable results in the clinical setting [27]. Our research team performed a study to evaluate the influence of image quality on the inter-rater agreement and reliability of the ICDSS scale and its impact on diabetic retinopathy screening assessment [26]. We concluded that the agreement between ophthalmologists in classifying images for quality is insufficient and affects the reliability of diabetic retinopathy screening results [26]. We also found that good or excellent image quality improved inter-rater reliability for the identification of diabetic retinopathy (with the κ value rising from 0.49 to 0.62 and the proportion of agreement increasing from 50% to 64%) [26]. Additionally, we reported that images classified as bad or fair quality were more likely to be classified as positive for diabetic retinopathy, referable diabetic retinopathy, and diabetic macular oedema, suggesting that inadequate image quality increases screen-positives [26]. This may be because the poor image quality that reduces the ophthalmologists' certainty to exclude disease; therefore, the graders tended to be more cautious and graded such images as positive, so that the patient would be referred. Consequently, more burden is put on screening programmes, increasing the number of patients requiring further observation. Only two ophthalmologists were included in that study, one being a retinal specialist and the other, a general ophthalmologist. This may limit the generalizability of the findings, but it also underscores the need for further studies on this topic.

It is worth noting the results for the reliability of image quality classification in the previous study by our research team [26]. With a kappa of 0.27 for image quality classification [26], these results suggest that the scale for image quality adopted may not be adequate for discriminating the quality level of fundus images, reinforcing the need to create reliable scales for image quality assessment. Indeed, the percent agreement for image quality reported in that study was 58%, meaning a lack of clear discrimination between the level of quality of a fundus image for providing a diabetic retinopathy screening result. This resulted in subjectivity in the classification of diabetic retinopathy, as other authors have previously reported [28]. Although, as far as we know, no other study has estimated the influence of image quality in diabetic retinopathy screening results provided by ophthalmologists, the study of Fu *et al* demonstrated that image quality highly influences the performance of automated classification systems of diabetic retinopathy [29].

5.5 Implications for clinical practice

There is insufficient interobserver agreement in classifying diabetic retinopathy lesions. Ungradable or poor-quality images have the clinical implication of wasting ophthalmologists' time in attempting to classify and interpret them. Furthermore, screen-positives lead to increased referrals for patient examination in the clinic by an ophthalmologist. The problem is being addressed: several automated screening systems use artificial intelligence (AI) algorithms for image quality assessment [30]. These algorithms are based on parameters such as colour, focus, contrast, illumination, and the presence of important structures (e.g., vessels, macula, optic disc, field), preventing unsuitable retinal images from being processed [30]. Furthermore, research has been conducted in the field of embedded image quality software in either hand-held fundus cameras or conventional fundus cameras [30, 31]. If these algorithms were integrated into the telemedicine diabetic retinopathy imaging pathway for screening in real-time, images would be immediately retaken, reducing the burden on ophthalmologists to exclude those of poor quality from being assessed for diabetic retinopathy [31]. This is particularly important considering the increasing number of diabetes patients to screen in the coming years [32] and their reduced adherence to diabetic retinopathy screening [33], meaning that the need for an additional visit to retake images would create a loss of opportunity to identify and treat diabetic retinopathy. Moreover, as suggested by our previous study [26], ensuring that only good or excellent quality images are sent to ophthalmologists may improve reliability for disease exclusion to 0.62 [26]. Also, in the study of Ruamviboonsuk *et al* [11], in which only good-quality images were used, the reliability of classification by retina specialists ranged between 0.63 and 0.67 [11].

As the results of our previous study demonstrated [26], it is difficult to establish guidelines for retinal image quality assessment to be used by ophthalmologists or AI algorithms for automated assessment of image quality, since ophthalmologists have limited agreement about the quality of fundus images for clinical decision-making in diabetic retinopathy screening. Therefore, developing consistent guidelines for image quality assessment and incorporating these algorithms into screening programmes may be essential for providing ophthalmologists with images of good or excellent quality and help establish a reliable diagnosis.

5.6 Conclusion

Although advances in technology and imaging techniques have greatly improved the quality of images, ophthalmologists still do not agree on whether an image is of sufficient quality for them to exclude diabetic retinopathy. Developing ways to increase their confidence in the image will require a better understanding of the underlying criteria for an image to be satisfactory. Furthermore, image quality influences the agreement and reliability of diabetic retinopathy classification. These aspects should be considered in the assessment of the quality of diabetic retinopathy screening programmes. A larger study is indicated to assess the interobserver variability with real images of different quality and define objective and reliable fundus image quality criteria for diabetic retinopathy classification based on ophthalmologists' image perception. Moreover, future studies should assess the real benefit of integrating automated image quality verification in clinical settings to relieve ophthalmologists from grading all images acquired.

References

[1] Mahabadi N and Al Khalili Y 2021 Neuroanatomy, retina *StatPearls* (Treasure Island, FL: StatPearls)

[2] Abràmoff M D, Garvin M K and Sonka M 2010 Retinal imaging and image analysis *IEEE Rev. Biomed. Eng.* **3** 169–208

[3] Kollias A N and Ulbig M W 2010 Diabetic retinopathy *Dtsch Arzteblatt Int.* **107** 75–84

[4] Chua J, Lim C X Y, Wong T Y and Sabanayagam C 2018 Diabetic retinopathy in the Asia-Pacific *Asia-Pac. J. Ophthalmol.* **7** 3–16

[5] Pieczynski J and Grzybowski A 2015 Review of diabetic retinopathy screening methods and programmes adopted in different parts of the world *Eur. Ophthal. Rev.* **9** 49

[6] Duh E J, Sun J K and Stitt A W 2017 Diabetic retinopathy: current understanding, mechanisms, and treatment strategies *JCI Insight* **2** e93751

[7] Wang W and Lo A C Y 2018 Diabetic retinopathy: pathophysiology and treatments *Int. J. Mol. Sci.* **20** 1816

[8] World Health Organization 2006 Prevention of blindness from diabetes mellitus - Report of a WHO consultation in Geneva, Switzerland, 9-11 November 2005 https://www.who.int/

[9] Yau J W Y *et al* 2012 Global prevalence and major risk factors of diabetic retinopathy *Diabetes Care* **35** 556–64

[10] Wu L, Fernandez-Loaiza P, Sauma J, Hernandez-Bogantes E and Masis M 2013 Classification of diabetic retinopathy and diabetic macular edema *World J. Diabetes* **4** 290–4

[11] Ruamviboonsuk P, Teerasuwanajak K, Tiensuwan M and Yuttitham K 2006 Interobserver agreement in the interpretation of single-field digital fundus images for diabetic retinopathy screening *Ophthalmology* **113** 826–32

[12] Gegundez-Arias M E, Ortega C, Garrido J, Ponte B, Alvarez F and Marin D 2016 Inter-observer reliability and agreement study on early diagnosis of diabetic retinopathy and diabetic macular edema risk *Bioinformatics and Biomedical Engineering. IWBBIO 2016* ed F Ortuño and I Rojas (Lecture Notes in Computer Science vol 9656) (Cham: Springer) pp 369–79

[13] Taylor D J *et al* 2009 Image-quality standardization for diabetic retinopathy screening *Expert Rev. Ophthalmol.* **4** 469–76

[14] Saine P J and Tyler M E 2002 Fundus photography overview *Ophthalmic Photography: Retinal Photography, Angiography, and Electronic Imaging* 2nd edn (Oxford: Butterworth-Heinemann) https://opsweb.org/page/fundusphotography

[15] Stewart M W 2016 Treatment of diabetic retinopathy: recent advances and unresolved challenges *World J. Diabetes* **7** 333–41

[16] Garg S and Davis R M 2009 Diabetic retinopathy screening update *Clin. Diabetes* **27** 140–5

[17] Chalakkal R J, Abdulla W H and Hong S C 2020 Fundus retinal image analyses for screening and diagnosing diabetic retinopathy, macular edema, and glaucoma disorders *Diabetes and Fundus OCT* ed A S El-Baz and J S Suri (Amsterdam: Elsevier) pp 59–111 ch 3

[18] http://sprawls.org/ppmi2/IMGCHAR/ 2018

[19] Cuadros J and Sim I 2004 EyePACS: an open source clinical communication system for eye care *Stud. Health Technol. Inform.* **107** 207–11

[20] Krupinski E A 2010 Current perspectives in medical image perception *Atten. Percept. Psychophys.* **72** 1205–17

[21] Kottner J and Streiner D L 2011 The difference between reliability and agreement *J. Clin. Epidemiol.* **64** 701–2

[22] Kottner J *et al* 2011 Guidelines for reporting reliability and agreement studies (GRRAS) were proposed *J. Clin. Epidemiol.* **64** 96–106

[23] McHugh M L 2012 Interrater reliability: the kappa statistic *Biochem. Medica* **22** 276–82

[24] Hallgren K A 2012 Computing inter-rater reliability for observational data: an overview and tutorial *Tutor. Quant. Methods Psychol.* **8** 23–34

[25] de Vet H C W, Terwee C B, Knol D L and Bouter L M 2006 When to use agreement versus reliability measures *J. Clin. Epidemiol.* **59** 1033–9

[26] Rêgo S, Dutra-Medeiros M, Bacelar-Silva G M, Borges T, Soares F and Monteiro-Soares M 2021 Reliability of classification by ophthalmologists with telescreening fundus images for diabetic retinopathy and image quality *J. Diabetes Sci. Technol.* **15** 710–2

[27] Fritz J M and Wainner R S 2001 Examining diagnostic tests: an evidence-based perspective *Phys. Ther.* **81** 1546–64

[28] Ma J J *et al* 2020 Diagnostic image quality assessment and classification in medical imaging: opportunities and challenges *Proc. IEEE Int. Symp. Biomed. Imaging* pp 337–40

[29] Fu H *et al* 2019 Evaluation of retinal image quality assessment networks in different color-spaces *Medical Image Computing and Computer Assisted Intervention Lecture Notes in Computer Science* vol 11764 (Berlin: Springer) pp 48–56

[30] Lin J, Yu L, Weng Q and Zheng X 2020 Retinal image quality assessment for diabetic retinopathy screening: a survey *Multimed. Tools Appl.* **79** 16173–99

[31] Teikari P, Najjar R P, Schmetterer L and Milea D 2019 Embedded deep learning in ophthalmology: making ophthalmic imaging smarter *Ther. Adv. Ophthalmol.* **11**

[32] Guariguata L, Whiting D R, Hambleton I, Beagley J, Linnenkamp U and Shaw J E 2014 Global estimates of diabetes prevalence for 2013 and projections for 2035 *Diabetes Res. Clin. Pract.* **103** 137–49

[33] Lewis K 2015 Improving patient compliance with diabetic retinopathy screening and treatment *Community Eye Health* **28** 68–9

IOP Publishing

Photo Acoustic and Optical Coherence Tomography Imaging, Volume 2

Fundus imaging for the retina

Ayman El-Baz and Jasjit S Suri

Chapter 6

Computer-aided diagnosis of Plus disease in preterm infants

Sivakumar Ramachandran, Bejoy Abraham, P Niyas, Anand Vinekar and Renu John

6.1 Introduction

Parts of this chapter are reprinted from [46], copyright (2021), with permission from Elsevier.

Medical image processing is an interdisciplinary area of research that attracts experts from various fields, namely physics, applied mathematics, engineering, statistics, biology, and medicine [1]. The prime intention of medical image processing is to assist clinicians in the effective diagnosis of various diseases. The research components consist of a number of stages starting from image acquisition and include other phases, namely image formation and reconstruction, image enhancement, image compression and storage, image analysis, visualization of the processed image, and finally evaluation and assessment of the obtained results [2]. This chapter focuses on infant retinal image processing and analysis for the early diagnosis of retinopathy of prematurity (ROP) in premature infants through Plus disease detection.

6.1.1 Retinal image analysis

The human eye is a well-built sensor that can assemble and process visually related intelligence from the environment. The retina inside the eye has the potential to serve as a window into the health of an individual. A non-invasive view of the retina with details of blood vessels and morphological structures could provide vital information that would enable physicians to make early detection and accurate diagnosis of many diseases. Studies established that changes in the retinal vascular structure can even provide indications of brain disorders [3, 4], in particular stroke [5] and vascular dementia [6]. The literature reports rapid development in the field of ophthalmic image processing techniques for eye disorders such as age-related macular degeneration (AMD), diabetic retinopathy (DR), glaucoma, and ROP [7].

The fundus region represents the posterior pole of the eye, including the morphological structures, namely retina, optic disc (OD), macula, fovea, and other interior vascular structures. Fundus examination facilitates early detection and diagnosis of many systemic and sight-threatening diseases [8]. The various ophthalmic imaging modalities such as optical coherence tomography (OCT), digital fundus photography, fluorescein angiography, stereo bio-microscopy, and indirect ophthalmoscopy are employed in screening of various eye diseases [9]. The fundus camera provides a cheap, simple, and easily accessible means to obtain a 2D representation of the 3D retinal structures among these different modalities. The fundus camera captures color images of the fundus that help to screen for the presence of diseases and monitor their changes over time.

Retinal diseases can be broadly classified into two categories, namely, diseases that are caused by vascular changes and disorders due to neuronal changes in the retina. A few diseases in both these categories include DR, cerebral white-matter lesion and atrophy, hypertension, nephropathy, congestive heart failure, stroke, Crohn's disease, glaucoma, myopia, coronary heart disease, Parkinson's disease, and ROP [10–16]. As this chapter focuses on computer-aided image analysis of preterm infant images, the discussion is limited to ROP and Plus disease.

6.1.1.1 Retinopathy of prematurity

ROP is a major cause of preventable childhood blindness across the world. The disease may lead to permanent blindness if not diagnosed and properly treated in its initial stage. ROP often co-occurs with Plus disease, a retinal disorder characterized by abnormal tortuosity and dilation of blood vessels [17]. The prognostic significance of Plus disease includes deciding the progress, stage, and severity of ROP in the affected retina of preterm infants. The clinicians often recommend initial stage ROP screening for premature babies with low birth weight and who are born before 31 weeks of gestation [18].

Recent advancements in neonatal care have increased the survival rate of low birth weight preterm infants. This improvement in survival rate has lead to an increase in ROP affected infants. Global statistics show that 50 000 preterm infants are affected by this disease every year, and in India the estimated vision loss is over 500 infants per year [19]. Thus, an effective screening protocol is essential for timely diagnosis and further treatment of this sight-threatening disease. The development of ROP in an infant retina is due to many complex factors. The initial stage of eye development occurs at about 16 weeks of pregnancy, when the blood vessels start to grow at the optic nerve head region. The newly formed blood vessels feed the developing retina with oxygen and nutrients, and they grow towards its edges. For a premature infant, vessels in the retina will not be fully developed, and they may not have reached the edges. Consequently, the peripheral retinal regions do not receive enough oxygen and nutrients, which simulate the production of vascular endothelial growth factor (VEGF) inside. This provides room for growing new abnormal vessels in the peripheral regions of the retina. The newly formed blood vessels are weak, fragile, and unstable. Retinal scarring occurs due to the rupture of newly formed blood vessels, and when these scars shrink, the retina is pulled away from the back of the eye leading to blindness. The presence of VEGF in the ischemic regions of the peripheral retina causes some interesting morphological changes in the blood vessels. These changes in retinal vessels are not entirely due to VEGF but also are accounted for by the blood flow mechanics inside the vessels [20, 21]. As a result, the arteries become tortuous, and veins become dilated due to their distensible property [22]. These vascular changes that occur in the posterior pole of the retina are known as Plus disease.

The International Committee for the Classification of ROP (ICROP), reconstituted in the year 2005, gave directions to ophthalmologists that the treatment of ROP should be initiated based on the detection of Plus disease [17]. The amount of abnormality needed to confirm a Plus case is obtained by comparing the image under consideration with a gold standard image [23]. As per recommendations made by the ICROP, vessel tortuosity greater than a particular threshold in any of two quadrants of the retina is a manifestation of Plus disease. The Committee also recommended a stage called pre-Plus for cases that cannot be considered as Plus but still have more venous dilation and arterial tortuosity than normal [24, 25]. Moreover, early diagnosis and treatment initiation achieve better results than treating the disease at some specific threshold points.

6.1.1.2 Infant retina imaging using Retcam

Binocular indirect ophthalmoscopy (BIO) forms the gold standard technique for examining an infant retina [26]. The diagnosis performed using BIO is based on drawings prepared by an expert after the examination is complete. The main drawbacks of this diagnostic system are: (i) the disease diagnosis depends on the mindset of the retinal expert, and there is a chance for incorrect decision as the drawings are made by recollecting from memory; (ii) the examination procedures make infants stressed; and (iii) there is a scarcity of retinal experts, particularly in rural areas [27]. Recently, Retcam has emerged as a better alternative for testing infant retina and ROP screening [26]. As telemedicine initiatives are becoming widespread, the use of Retcam, which captures digital images of the retina, ensures effective storage, diagnosis, and management of the disease. In India, the Karnataka Internet Assisted Diagnosis of Retinopathy of Prematurity (KIDROP) is such an initiative managed by Narayana Netralaya [28], Bangalore, in collaboration with the National Rural Health Mission, the Government of India, and the Ministry of Health and Family Welfare, Government of Karnataka, India, to tackle ROP in rural areas of the country. The images used for this study were obtained from KIDROP, Bangalore. In figure 6.1 examples of normal and Plus disease affected retinal images obtained from KIDROP are shown for illustration.

6.2 Computer-aided diagnosis system

The conventional diagnosis procedure to detect Plus disease for ROP diagnosis involves an expert comparing the tortuous vessel segments present in an image with that of a gold standard image. The disease prediction depends solely on the

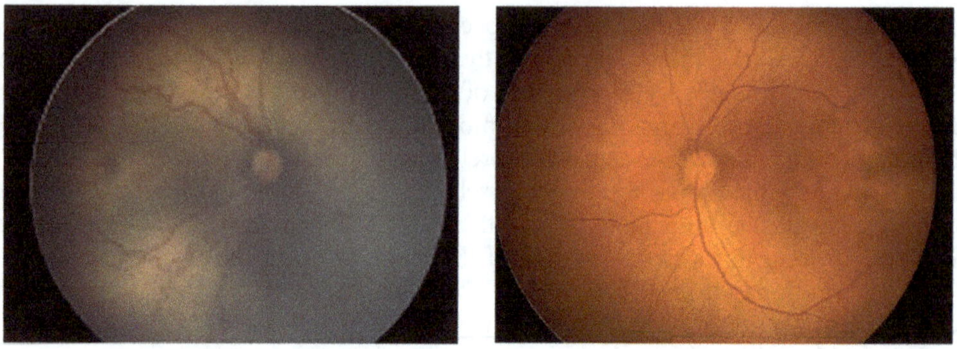

(a) (b)

Figure 6.1. Retinal images of premature infants acquired using the Retcam fundus camera. (a) Cases of Plus disease (showing signs of ROP) and (b) normal healthy images.

experience as well as the mindset of the inspecting expert. Moreover, the manual examination process is a time consuming as well as a tedious task. As telemedicine-based diagnosis is becoming widespread, the number of images that need to be screened are too high, and manual examination of these images becomes very difficult. Hence a feasible solution is in high demand. Computer-aided diagnosis (CAD) that employs computers for disease prediction and further analysis provides a better alternative to this problem. The end goal of CAD is not automated diagnosis but an improved diagnostic decision by the physician using quantitative and objective analysis facilitated by the application of computers in medical images.

A substantial amount of research works have been proposed for the CAD of Plus disease. Most of the works use the tortuosity of blood vessels as a major parameter for disease prediction. A few amount works compute the width of retinal blood vessels as an additional parameter along with tortuosity for disease diagnosis. The research works that use vessel parameters for tortuosity evaluation either use manually segmented or automatically detected blood vessels in a predefined area around the OD. Recently researchers used deep learning networks to classify a given image as diseased or healthy. In this chapter, the discussion is limited to those techniques that employ deep learning for disease prediction.

6.2.1 Resources and evaluation metrics

The study population is composed of retinal images of premature infants with an average birth weight of 2000 grams and/or gestational age fewer than 34 weeks, acquired during routine ROP screening procedures performed at Narayana Netralaya, Bangalore, India under a program called KIDROP. KIDROP is the world's largest telemedicine network, which aims to eradicate ROP-based infant blindness. So far no research groups have made their image data available publicly for testing and validation of newly developed techniques for Plus disease diagnosis.

The KIDROP dataset contains 289 retinal images of preterm infants that belong to the category of disc-centered images, where the OD lies in the image center. The

Table 6.1. The statistics of infant subjects of the KIDROP dataset. The image set consists of 200 healthy and 89 Plus-diseased subjects with mean and standard deviation (SD) as shown. Here, BW, GA, and PM are, respectively, birth weight in grams (g), gestational age in weeks (w), and post-menstrual age in weeks (w).

	BW (mean ± SD)	GA (mean ± SD)	PM (mean ± SD)
No Plus	1350 ± 240 g	32.4 ± 1.1 w	38.5 ± 2.3 w
Plus	1280 ± 226 g	30.9 ± 1.8 w	36.5 ± 1.9 w

images were captured using a RetCam3™ camera with a field of view of 130 degrees and a resolution of 1600 × 1200 pixels. Of the 289 images, 89 contain signs of Plus disease, while the remaining were marked as healthy. The images were captured by skilled technicians with ample experience in eye-screening. For this study, a maximum of two images, one from each eye, were selected from a patient and the whole dataset corresponds to retinal images of 150 premature infants. Hence there exists wide variability among the images in the used dataset. The images were graded as 'Plus disease' or 'no Plus disease' by three independent ROP specialists at different times and the responses were recorded. Only images for which all three responses were identical are included for analysis. This was performed to reduce the variability that is well known in the grading of Plus disease, even by experts, as reported in the literature [30]. In table 6.1 we provide the details of birth weight (BW), gestational age (GA), and post-menstrual age (PM) of the imaged subjects.

6.2.1.1 Performance measures

Here, four kinds of outcomes can be defined when the diagnosis of the disease is made by classifying images as diseased or healthy using a developed algorithm. When an image labeled as diseased by clinicians is correctly classified, a true positive (T1) is registered and a false negative (F2) when incorrectly classified. Similarly, if an image of a healthy retina is correctly classified a true negative (T2) exists and a false positive (F1) if it is categorized as abnormal. In the disease diagnosis, the presence of disease in an image is marked as logic 1 and 0 represents a healthy case. To quantify the output results, a few performance measures are computed using the four outcomes. The true positive rate (TPR) or sensitivity (Se) is expressed as the fraction of actual positive cases that are identified correctly. It represents the capability of the proposed system to detect true positives. True negative rate (TNR) or specificity (Sp) represents the ability to detect true negative cases. Using TI, T2, F1, and F2, the following performance measures are evaluated:

$$\begin{aligned} \mathrm{Se} &= \frac{\mathrm{T1}}{\mathrm{T1}+\mathrm{F2}}, \quad \mathrm{Sp} = \frac{\mathrm{T2}}{\mathrm{T2}+\mathrm{F1}}, \quad \mathrm{Pr} = \frac{\mathrm{T1}}{\mathrm{T1}+\mathrm{F1}}, \quad \mathrm{F} = 2\frac{\mathrm{Pr.\,Se}}{\mathrm{Pr}+\mathrm{Se}}, \quad \text{and} \\ \mathrm{Acc} &= \frac{\mathrm{T1}+\mathrm{T2}}{\mathrm{T1}+\mathrm{T2}+\mathrm{F1}+\mathrm{F2}}, \end{aligned} \tag{6.1}$$

where Pr, F, and Acc represent precision, the *F*-measure, and accuracy, respectively. In addition to the above performance metrics, the Matthews correlation coefficient (MCC) and Cohen's kappa coefficient (κ) are commonly used performance measures of unbalanced classes in binary classification problems. The MCC is formally defined as

$$\mathrm{MCC} = \frac{\mathrm{T1.\ T2} - \mathrm{F1.\ F2}}{\sqrt{\mathrm{T1} + \mathrm{F1}}(\mathrm{T1} + \mathrm{F2})(\mathrm{T2} + \mathrm{F1})(\mathrm{T2} + \mathrm{F2})}. \tag{6.2}$$

Its values range between +1 and −1, where +1 indicates a perfect prediction, 0 indicates a random prediction and −1 corresponds to opposite predictions. The kappa coefficient (κ) also deals with imbalanced class problems and is defined as

$$\kappa = \frac{p_{\mathrm{obs}} - p_{\mathrm{exp}}}{1 - p_{\mathrm{exp}}}, \tag{6.3}$$

where p_{obs} is the observed agreement and p_{exp} is the expected agreement. The κ values vary between −1 and +1, with values less than or equal to 0 indicating no agreement, and values greater than 0 indicating the degree of agreement, with a maximum value equal to 1.

6.3 Development of a computer-aided diagnostic system

The diagnosis of Plus disease in a given infant retinal image is based on the presence of tortuous vessels in the images. The disease diagnosis that relies on conventional image processing based techniques initially segments the retinal blood vessels present in an image followed by tortuosity analysis in the segmented vessel map. Recently, deep learning (DL) has greatly affected many computer vision tasks such as image segmentation, object localization and tracking, image classification, object detection, pose estimation, image captioning, etc [31]. The advancements in the development of powerful graphical processing units (GPUs), an increase in the availability of labeled data, and sharing of source code along with developed models among the research community have paved the way for the widespread use of deep neural networks (DNN) in research.

Until recently, researchers have reported only a few studies based on DNN for the early diagnosis of ROP. The main reason for this scarcity is the unavailability of ROP images for experimental evaluation. Moreover, the Retcam images are too costly which prevents research groups from making their experimental data available publicly. The research works published so far have used a large amount of training datasets to classify a given image as diseased or normal. The network extracts the image features and predicts the existence of ROP in the image. The ROP-DL system proposed in [32] used a two-stage procedure for ROP prediction. In the first stage retinal vessels were segmented using a U-Net architecture trained on a very large number of image patches of size 48 × 48 pixels extracted from manually labeled vessel images. The Inception NN architecture [33] was then used to classify retinal images as a Plus, pre-Plus, or normal cases. The work introduced in [34] employed two specific DNN models for ROP prediction. The detection comprises two sub-tasks, namely classification and grading. The former distinguished images with ROP cases from normal ones, while the latter graded ROP images into different levels based on the severity of the disease. To deal with the tasks, an ROP detection algorithm called DeepROP was developed using the Inception NN architecture. A dropout trained CNN-based ROP detection system using a pre-trained GoogLeNet [35] combined with the Bayesian framework was proposed in [36] and claimed high accuracy in prediction. In that work, a second CNN was trained to visualize the feature maps of pathologies that occur due to the attack of the disease. In [37] a labeled training set fine-tuned three DNN classifiers with a transfer learning approach for the classification of images as normal or diseased. The dataset used consisted of 420 365 wide-angle infant retina images collected from various sources. A pre-processing step was used to refine the images before training and performance evaluation. The training images fine-tuned three DNN classifiers, namely AlexNet [38], GoogLeNet, and VGG-16 [39]. The resultant models were evaluated on a test image set of 1742 samples and compared with the predictions of pediatric ophthalmologists. In the work proposed by Brown *et al* [32], the U-Net architecture was tuned to segment out vessels from the infant retinal images. The images were classified into pre-Plus, Plus disease, or normal from the segmented vessel map using the Inception architecture. A graded classification of fundus images was performed in [40] to classify images into Plus or pre-Plus categories using the Inception-v3 architecture. The RMSProp optimizer with binary cross-entropy (BCE) loss was used for training the network. The probabilistic outputs were generated from the softmax layer placed at the output section.

Although DNN-based approaches are quite promising, the interpretation of output results needs expert intervention, making the process time consuming. This limits its use in real medical problems, where a simple and understandable explanation of the diagnostic decision has to be provided. In this context, here we discuss a deep learning model for ROP diagnosis via Plus disease prediction. The framework is straightforward, explainable, and has many advantages compared to the existing methods. The outputs obtained can be interpreted easily by an ophthalmologist as it highlights pathological regions, which are the tortuous vessel points present in the images. The performance of the proposed pipeline is validated on the dataset of 289 images, obtained from KIDROP Bangalore, India.

6.3.1 Tortuosity detection and quantification using a fully convolutional neural network

In this section the discussion is limited to a DL framework for detecting and quantifying tortuous vessel segments, leading to effective Plus disease diagnosis in preterm infants. The network considers tortuosity as a regression problem, where abnormal vessel locations present in an infant retinal image are directly converted to bounding first box dimensions. The method employs a fully CNN architecture for detecting tortuous vessel segments in the given retinal image. The architecture of the proposed system is adapted from the YOLO (You Look Only Once) object detection network [41]. The results obtained using the proposed system are similar to those shown in [29], where a novel U-COSFIRE filter detects tortuous locations in an infant retinal image.

6.3.1.1 Methodology

Figure 6.2 shows a schematic block diagram of the pipeline for the detection and quantification of tortuous vessel segments present in preterm infant images. Various operations performed in the methodology are discussed below.

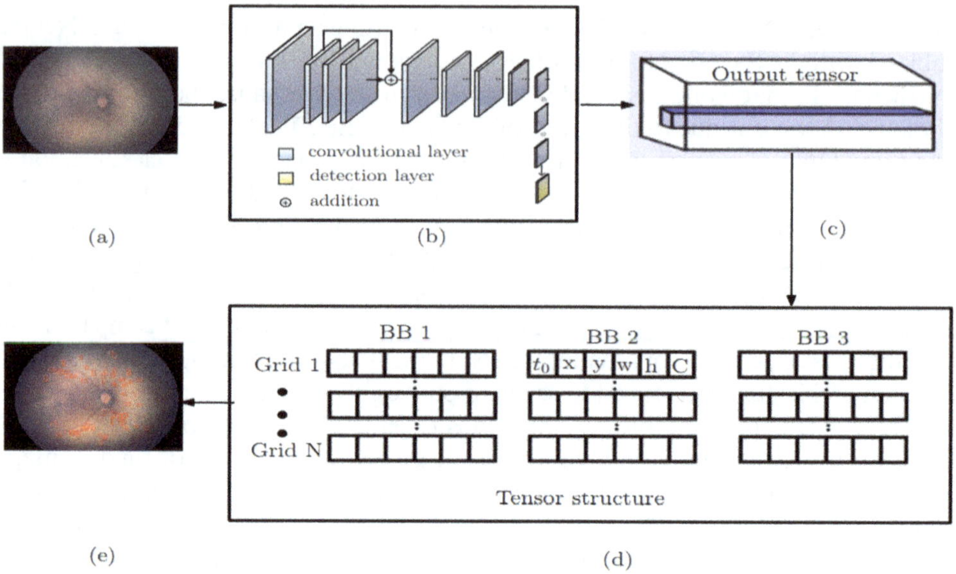

Figure 6.2. Overview of the proposed approach. (a) Input image, (b) schematic of the proposed network structure, (c) output tensor, (d) tensor structure of dimension $N \times N \times 18$ ($N = 52$), and (e) output image with marked tortuous vessel segments. BB denotes the bounding box.

6.3.1.1.1 Data augmentation

Deep learning-based medical research works are often constrained to the availability of image data as a large number of images are required for training the network for achieving high accuracy. The overall performance of deep neural networks improves with the amount of training data. To some extent, image data augmentation can be performed to fix this problem by adding transformed images to the training samples. The process involves generating perturbed versions of training images using the same class of existing images in a given dataset. Commonly employed augmentation methods include scaling, flipping, rotation, translation, noise addition, and intensity shifts. The augmentation technique intends to enlarge the training samples with new convincing examples suited for a particular job at hand. This means that the newly added samples should be variations of the training data that are likely to come across into the model at different instances. Moreover, there is no need to add those data, which the model has the least chance to encounter. Thus the choice of a particular data augmentation technique employed for training a dataset must be chosen carefully based on the knowledge of the problem domain. For the reported work, the size of the training data is expanded by flipping images horizontally, and color intensity shifts in HSV color space.

6.3.1.1.2 Network architecture

The network architecture consists of 33 convolutional layers with variable size filters, activation functions, batch normalization, and residual layers. The role of a convolutional layer is to extract feature maps from the entire retinal image using learnable filters of kernel size, W. The ith feature map of Kth convolutional layer, Y_K^i is evaluated as

$$Y_K^i = f(W_K^i * Y_{K-1}^i), \tag{6.4}$$

where $f(\cdot)$ represents activation function and $*$ denotes convolution operation. The network employs a leaky rectified linear unit (ReLU) activation function in all layers, which prevents the saturation of positive as well as negative inputs. For an input x_i, the ReLU function is defined as

$$f(x_i) = \begin{cases} x_i, & x_i \geqslant 0 \\ \propto x_i, & \text{otherwise} \end{cases}. \tag{6.5}$$

The negative slope coefficient α is set as 0.1. The feature maps are down-sampled by changing the stride of the network layers, which help to select proper features and also in size reduction of the feature maps. The network employs residual blocks [42] that assist the generation of better features by adding old features using skip connection to a learned residue. The network weights are normalized by using a technique called batch normalization [43], which reduces the variance between layers and accelerates the training of the model. The features extracted from a layer is linearly transformed before it is applied to the succeeding layers. The network weights are updated just before it is applied to the activation function. The network splits a given input image into 52 × 52 grid cells and predicts three bounding boxes per cell. Each box is described by five attributes, namely t_0, x, y, w, and h. Here t_0 is a confidence measure that indicates the presence of a tortuous vessel segment inside the predicted box and is evaluated using logistic regression. The confidence t_0 is defined as the product of the probability that a grid cell contains a tortuous vessel segment (P_{tv}) and intersection of union (IoU) between the predicted bounding box and the ground truth box:

$$t_0 = P_{\text{tv}} \times \text{IoU (predicted, ground truth)}. \tag{6.6}$$

The confidence metric is zero if there is no tortuous vessel segment present in a particular grid cell. The grid cell responsible for the detection of an abnormal vessel segment will have its centroid lying in that cell. Bounding boxes with a lower probability of tortuous vessel segment in a grid cell are removed and those having the highest IoU with the ground truth are retained. The coordinates (x, y) represent the center of the bounding box, whose width and height are represented as w and h, respectively. The input image is resized to 416 × 416 pixels before it is applied to the network. The image is then down-sampled in successive stages to obtain a feature map of size 52 × 52. The network output is a tensor of dimension $52 \times 52 \times (B \times (B_a + C))$, where B represents the number of bounding boxes predicted per cell, B_a represents the bounding box attributes and C, the object class that is the tortuous vessel segments present in the image. The final detection is made using a filter of size 1 × 1, resulting in an output dimension 52 × 52 × 18 (here $B = 3$, $B_a = 5$, and $C = 1$). The detailed architecture of the network is shown in table 6.2.

Table 6.2. The model architecture used for tortuous vessel detection.

	Type	Filter	Size/Stride	Output
1×	Convolutional	32	3 × 3	416 × 416 × 32
	Convolutional	64	3 × 3/2	208 × 208 × 64
1×	Convolutional	32	1 × 1	208 × 208 × 32
	Convolutional	64	3 × 3	208 × 208 × 64
	Residual			
1×	Convolutional	128	3 × 3/2	104 × 104 × 128
2×	Convolutional	64 128	1 × 1	104 × 104 × 64
	Convolutional		3 × 3	104 × 104 × 128
	Residual			
1×	Convolutional	256	3 × 3/2	52 × 52 × 256
8×	Convolutional	128	1 × 1	52 × 52 × 128
	Convolutional	256	3 × 3	52 × 52 × 256
	Residual			
3×	Convolutional	128	1 × 1	52 × 52 × 128
	Convolutional	256	3 × 3	52 × 52 × 256
	Convolutional	18	1 × 1	52 × 52 × 18

The bounding box prediction in detection tasks can be accomplished in different ways. One approach is the direct prediction of box parameters. However, this procedure may lead to instability in gradients while training the network. Alternatively, the second approach uses bounding box priors called anchor boxes and predicts offset to these predefined boxes. These box priors are set based on the object sizes in training data, which capture the scale and aspect ratio of different tortuous vessel segments to be detected in the image. During detection, the anchor boxes are tiled across the image and the network predicts offsets to these prior boxes. The bounding box responsible for the detection of tortuous vessel segments will have the highest IoU between its anchor and the ground truth box.

Good candidate anchor boxes are computed from training images using k-means clustering and corrections are applied to these prior boxes to match the dimensions of the ground truth bounding box. The distance measure between clusters is evaluated using the IoU metric with the width and height of the bounding boxes as features.

6.3.1.1.3 Tortuosity detection and quantification

The test images given as input to the trained network are analysed to detect the tortuous vessel segments. The network output bounding boxes over tortuous vessel segments present in the image. The number of bounding boxes in each image, which corresponds to the number of tortuous vessel segments is counted and and used further to diagnose the disease. The total count of bounding boxes in each image is analysed for the diagnosis of Plus disease. A high value shows the abundance of tortuous vessels in the image and a positive diagnostic score for the presence of ROP.

6.4 Results and discussion

The network is initially trained with 288 labeled images obtained using augmentation of 32 manually labeled images from the dataset. The detection of tortuous vessel regions in the test images is performed based on a two-step procedure that suppresses non-maximal bounding boxes in the image. Each grid cell output a confidence score in the test image that serves as the class probability of the maximum confidence among the three predicted bounding boxes. The proposed methodology generates many potential regions of interest (ROIs) in each test image. A threshold is set to limit the generation of ROIs (tortuous vessel segments). The set threshold should not be high, as it causes loss of detectable tortuous vessel segments. On the other hand, very low threshold values lead to the detection of unwanted ROIs with less class probability. In the reported study, the set threshold value is 0.3, which achieves the loss of a lower number of detectable tortuous vessel segments. In the first step, the predicted bounding boxes with a lower detection probability than the set threshold are removed from the image. In the next step, an IoU threshold is set to select the bounding boxes with the highest detection probability and remove all the bounding boxes whose IoU value is greater than the set IoU threshold, which is equal to 0.45. Thus all predicted bounding boxes with an IoU value greater than 0.45 with respect to the best bounding box will be removed.

In figure 6.3 some instances of assessment performed on three images of the KIDROP dataset are shown for illustration. The number of detected bounding boxes in the retinal image with disease is 51 and 62, respectively (figures 6.3(a) and (b)), while that on the healthy image is 4 (figure 6.3.(c)). This shows how the proposed method can extract valuable information for evaluating the blood vessel tortuosity and thereby identify the presence of Plus disease for subsequent detection of ROP in the infant retinal images. The reported sensitivity and specificity are 0.99 and 0.98, respectively, for the maximum F-measure obtained (0.98). The MCC and κ values at this threshold level are 0.96. The high AUC equal to 0.9947 shows the capability of the model in distinguishing diseased and healthy images.

A comparison of results with the findings reported in recent studies employing CNN is also reported here. The results provided in table 6.3 are those reported using different datasets. Hence, instead of a direct comparison, a qualitative analysis of the results obtained with those of the existing ones is presented. Worrall *et al* [36] performed studies using a deep learning ROP detection framework consisting of fully connected layers stacked on top of GoogleNet. A dataset consisting of 1459 images was used in the study and performance metrics are obtained using nine-fold cross-validation. In [34] an automated ROP identification and grading system using two specific networks was designed using the Inception-v3 network. The dataset used consists of 20 795 images and the network was trained using a large number of images. The work proposed in [40] uses Inception-v3 to classify an image as healthy or

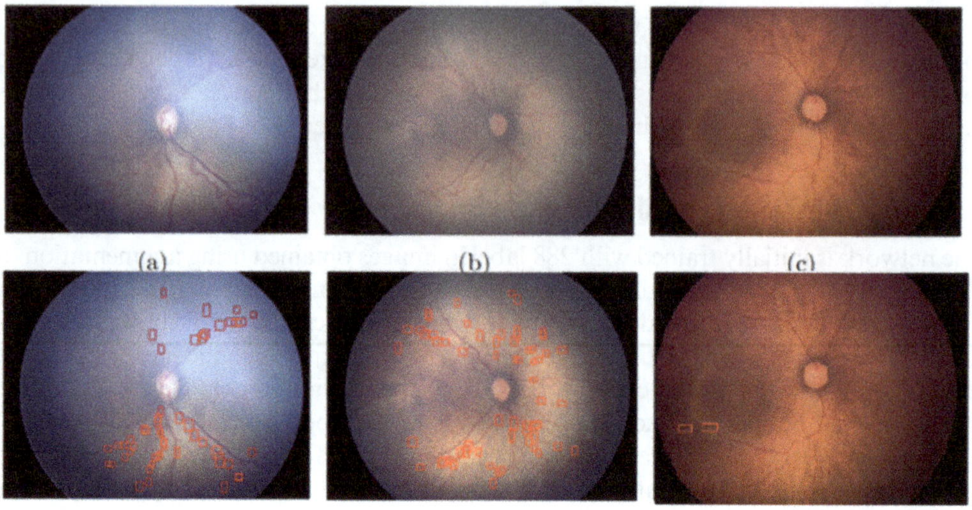

Figure 6.3. The output obtained for images N01, N18, and N28 of the KIDROP dataset. The top row shows the input images and the bottom row shows the image with tortuous locations marked on the vessels. Images (a) and (b) are diseased and (c) is healthy.

Table 6.3. Comparison of the results reported in similar DNN studies on ROP detection. The blank entries are those which are not specified in the original work.

Study	Data Type	Sensitivity	Specificity	Method
Worrall *et al* [36]	Private	0.83	0.98	CNN-based
Wang *et al* [34]	Private	0.97	0.99	CNN-based
Tan *et al* [40]	Private	0.96	0.98	CNN-based
Zhang *et al* [37]	Private	0.99	0.98	CNN-based
Wang *et al* [44]	Private	0.90	—	CNN-based
Peng *et al* [45]	Private	0.94	—	CNN-based
Reported work [46]	Private	0.99	0.98	CNN-based

diseased. The study is performed in a dataset of 6974 images obtained after augmenting 3487 original raw images. In a work proposed by Zhang *et al* [37], the performance of VGG-16 net, AlexNet, and GoogLeNet were compared with that of a human ROP expert and showed that the VGG-16 and human performances are comparable. Wang *et al* [44] proposed a DL-based ROP detection method using Inception modules with skip connections. The work employs a dataset consisting of 11 000 images and the imbalance present in the data is rectified using median frequency balancing. In a recent work proposed in [45], ROP classification is done with a pre-trained ResNet18 DNN trained using 8351 image samples. The proposed model used a CNN pre-trained on natural images and fine-tuned for ROP classification.

Compared with the limited DL-based ROP detection methods, the advantages of the reported pipeline are as follows. (a) The output results obtained mimic the diagnostic procedure performed by an expert in the sense that exact tortuous locations are accurately marked in images. The high explainability of the output makes it very useful for a clinician, who can incorporate his/her knowledge along with the obtained results for effective disease diagnosis. (b) The ROP detection in the existing DL-based methods depends on the amount of training data used to learn the model. The image is classified as diseased or healthy using a large amount of labeled image data and there is no inference about the attainability of the predicted results. In this method, markings are made over the tortuous vessel segments, allowing a visual interpretation of the results. (c) The performance of the pipeline relies not on the number of images used for training, but rather on the number of marked tortuous vessel segments present in the training images.

The reported DL-based diagnostic method has some limitations. Even though the results are explainable, the network is still a black box. The user is not aware of the internal operations of the network. The hidden layer visualization techniques may be explored for a better understanding of the network. The mere Plus-disease prediction is insufficient for diagnosing the disease in clinical situations as ROP diagnosis is a complex procedure and includes different stages, namely Plus and pre-Plus, and severity of the disease, such as AP-ROP, acute ROP, and pre-threshold ROP. In the future, improved models trained using detailed labeled images can be used to improve the performance of the system.

The ROP diagnostic techniques and the reviews presented in this chapter are based on deep neural networks alone. The methodology presented here is based on our research papers published in [46, 47]. It is equally worth noting that there exists a substantial amount of ROP research based on conventional image processing techniques [48–54]. The pace of research in ROP diagnosis using premature retinal images is hindered due to the unavailability of a public image dataset for validation of newly developing algorithms. Hence, there still exists a lack of full-fledged diagnostic systems that can be used as an adjunct tool in ROP diagnosis, which promises the future scope of this research.

6.5 Conclusions

In this chapter we discuss a conceptually simple and highly effective CAD system for the detection of Plus disease in retinal fundus images of preterm infants. Our study narrates the efficacy of the methodology that directly predicts the tortuous vessel locations in a given infant retinal image. The results obtained show the high effectiveness of the pipeline, as a computer-aided diagnostic tool, for the early detection of ROP through Plus disease prediction.

References

[1] Zhou H, Wu J and Zhang J 2010 *Digital Image Processing: Part II* (Copenhagen: Bookboon)

[2] Duncan J S and Ayache N 2000 Medical image analysis: progress over two decades and the challenges ahead *IEEE Trans. Pattern Anal. Mach. Intell.* **22** 85–106

[3] Tso M O and Jampol L M 1982 Pathophysiology of hypertensive retinopathy *Ophthalmology* **89** 1132–45

[4] Patton N, Aslam T, MacGillivray T, Pattie A, Deary I J and Dhillon B 2005 Retinal vascular image analysis as a potential screening tool for cerebrovascular disease: a rationale based on homology between cerebral and retinal microvasculatures *J. Anat.* **206** 319–48

[5] Sadek J R and Hammeke T A 2002 Functional neuroimaging in neurology and psychiatry *CNS Spectr.* **7** 286–99

[6] Yoshikawa T *et al* 2003 Heterogeneity of cerebral blood flow in Alzheimer disease and vascular dementia *Am. J. Neuroradiol.* **24** 1341–7

[7] Patton N, Aslam T M, MacGillivray T, Deary I J, Dhillon B, Eikelboom R H, Yogesan K and Constable I J 2006 Retinal image analysis: concepts, applications and potential *Progr. Retin. Eye Res.* **25** 99–127

[8] Herbert H, Jordan K and Flanagan D 2003 Is screening with digital imaging using one retinal view adequate? *Eye* **17** 497–500

[9] Ilginis T, Clarke J and Patel P J 2014 Ophthalmic imaging *Br. Med. Bull.* **111** 77–88

[10] Roychowdhury S, Koozekanani D D, Radwan S and Parhi K K 2013 Automated localization of cysts in diabetic macular edema using optical coherence tomography images *35th IEEE Annual Int. Conf. of the Engineering in Medicine and Biology Society* (Piscataway, NJ: IEEE) pp 1426–9

[11] Archibald N K, Clarke M P, Mosimann U P and Burn D J 2009 The retina in Parkinson's disease *Brain* **132** 1128–45

[12] Witt N, Wong T Y, Hughes A D, Chaturvedi N, Klein B E, Evans R, McNamara M, Thom S A M and Klein R 2006 Abnormalities of retinal microvascular structure and risk of mortality from ischemic heart disease and stroke *Hypertension* **47** 975–81

[13] Falavarjani K G, Parvaresh M M, Shahraki K, Nekoozadeh S and Amirfarhangi A 2012 Central retinal artery occlusion in Crohn disease *J. Am. Assoc. Pediat. Ophthalmol. Strabismus* **16** 392–3

[14] Muramatsu C, Hayashi Y, Sawada A, Hatanaka Y, Hara T, Yamamoto T and Fujita H 2010 Detection of retinal nerve fiber layer defects on retinal fundus images for early diagnosis of glaucoma *J. Biomed. Opt.* **15** 016021

[15] Wong T Y, Klein R, Klein B E, Tielsch J M, Hubbard L and Nieto F J 2001 Retinal microvascular abnormalities and their relationship with hypertension, cardiovascular disease, and mortality *Surv. Ophthalmol.* **46** 59–80

[16] Wong T Y, Shankar A, Klein R, Klein B E and Hubbard L D 2005 Retinal arteriolar narrowing, hypertension, and subsequent risk of diabetes mellitus *Arch. Intern. Med.* **165** 1060–5

[17] Gole G A *et al* 2005 The international classification of retinopathy of prematurity revisited *JAMA Ophthalmol.* **123** 991–9

[18] Blencowe H, Lawn J E, Vazquez T, Fielder A and Gilbert C 2013 Preterm-associated visual impairment and estimates of retinopathy of prematurity at regional and global levels for 2010 *Pediat. Res.* **74** 35–49

[19] Le C *et al* 2016 Retinopathy of prematurity: incidence, prevalence, risk factors, and outcomes at a tertiary care center in Telangana *J. Clin. Ophthalmol. Res.* **4** 119

[20] Smith L, Joussen A, Gardner T, Kirchhof B and Ryan S 2007 Retinopathy of prematurity: pathophysiology of disease *Retinal Vascular Disease* (Berlin/Heidelberg: Springer) 392–402

[21] Hartnett M E, Martiniuk D, Byfield G, Geisen P, Zeng G and Bautch V L 2008 Neutralizing VEGF decreases tortuosity and alters endothelial cell division orientation in arterioles and

veins in a rat model of ROP: relevance to Plus disease *Investig. Ophthalmol. Vis. Sci.* **49** 3107–14

[22] Johnston S C, Wallace D K, Freedman S F, Yanovitch T L and Zhao Z 2009 Tortuosity of arterioles and venules in quantifying Plus disease *J. Am. Assoc. Pediat. Ophthalmol. Strabismus* **13** 181–5

[23] Oloumi F, Rangayyan R M and Ells A L 2014 Quantification of the changes in the openness of the major temporal arcade in retinal fundus images of preterm infants with Plus disease *Investigat. Ophthalmol. Vis. Sci.* **55** 6728–35

[24] Hardy R, Good W, Dobson V, Palmer E, Tung B and Phelps DEarly Treatment for Retinopathy of Prematurity Cooperative Group Revised indications for the treatment of retinopathy of prematurity. Results of the early treatment for retinopathy of prematurity randomized trial *Arch. Ophthalmol.* **121** 1684–94

[25] Reynolds J D *et al* 2002 Evidence-based screening criteria for retinopathy of prematurity: natural history data from the CRYO-ROP and LIGHT-ROP studies *Arch. Ophthalmol.* **120** 1470–6

[26] Sekeroglu M, Hekimoglu E, Sekeroglu H and Arslan U 2013 Alternative methods for the screening of retinopathy of prematurity: binocular indirect ophthalmoscopy vs wide-field digital retinal imaging *Eye* **27** 1053–7

[27] Dhaliwal C, Wright E, Graham C, McIntosh N and Fleck B W 2009 Wide-field digital retinal imaging versus binocular indirect ophthalmoscopy for retinopathy of prematurity screening: a two-observer prospective, randomised comparison *Br. J. Ophthalmol.* **93** 355–9

[28] Vinekar A 2011 IT-enabled innovation to prevent infant blindness in rural India: the KIDROP experience *J. Indian Business Res.* **3** 98–102

[29] Ramachandran S, Strisciuglio N, Vinekar A, John R and Azzopardi G 2020 UCOSFIRE filters for vessel tortuosity quantification with application to automated diagnosis of retinopathy of prematurity *Neur. Comput. Appl.* **32** 12453–68 (https://link.springer.com/article/10.1007/s00521-019-04697-6)

[30] Gschließer A *et al* 2015 Inter-expert and intra-expert agreement on the diagnosis and treatment of retinopathy of prematurity *Am. J. Ophthalmol.* **160** 553–60

[31] Gu J *et al* 2018 Recent advances in convolutional neural networks *Pattern Recognit.* **77** 354–77

[32] Brown J M *et al* 2018 Automated diagnosis of Plus disease in retinopathy of prematurity using deep convolutional neural networks *JAMA Ophthalmol.* **136** 803–10

[33] Szegedy C, Ioffe S, Vanhoucke V and Alemi A A 2017 Inception-v4, InceptionResNet and the impact of residual connections on learning *31st AAAI Conf. on Artificial Intelligence*

[34] Wang R, Ju Y, Chen L, Zhang J, Hu Y, Wu W, Dong J, Zhong and Yi Z 2018 Automated retinopathy of prematurity screening using deep neural networks *EBioMed.* **35** 361–8

[35] Szegedy C, Liu W, Jia Y, Sermanet P, Reed S, Anguelov D, Erhan D, Vanhoucke V and Rabinovich A 2015 Going deeper with convolutions *Proc. IEEE Conf. on Computer Vision and Pattern Recognition* pp 1–9

[36] Worrall D E, Wilson C M and Brostow G J 2016 Automated retinopathy of prematurity case detection with convolutional neural networks *Int. Workshop on Large-Scale Annotation of Biomedical Data and Expert Label Synthesis* (Berlin: Springer) pp 68–76

[37] Zhang Y, Wang L, Wu Z, Zeng J, Chen Y, Tian R, Zhao J and Zhang G 2018 Development of an automated screening system for retinopathy of prematurity using a deep neural network for wide-angle retinal images *IEEE Access* **7** 10232–41

[38] Krizhevsky A, Sutskever I and Hinton G E 2012 ImageNet classification with deep convolutional neural networks *Commun. ACM* **60** 84–90

[39] Simonyan K and Zisserman A 2014 Very deep convolutional networks for large-scale image recognition arXiv: 1409.1556

[40] Tan Z, Simkin S, Lai C and Dai S 2019 Deep learning algorithm for automated diagnosis of retinopathy of prematurity Plus disease *Translat. Vision Sci. Technol.* **8** 23

[41] Redmon J and Farhadi A 2018 Yolov3: an incremental improvement arXiv:1804.02767

[42] He K, Zhang X, Ren S and Sun J 2016 Deep residual learning for image recognition *Proc. IEEE Conf. on Computer Vision and Pattern Recognition* pp 770–8

[43] Ioffe S and Szegedy C 2015 Batch normalization: accelerating deep network training by reducing internal covariate shift arXiv:1502.03167

[44] Wang Y and Chen Y 2019 Automated recognition of retinopathy of prematurity with deep neural networks *J. Phys. Conf. Ser.* **1187** 042057

[45] Peng Y, Zhu W, Chen F, Xiang D and Chen X 2020 Automated retinopathy of prematurity screening using deep neural network with attention mechanism *Medical Imaging 2020: Image Processing* vol 11313 (Bellingham, WA: International Society for Optics and Photonics) p 1131321

[46] Ramachandran S, Niyas P, Vinekar A and John R 2021 A deep learning framework for the detection of Plus disease in retinal fundus images of preterm infants *Biocybern. Biomed. Eng.* **41** 362–75

[47] Ramachandran S, Kochitty S, Vinekar A and John R 2020 A fully convolutional neural network approach for the localization of optic disc in retinopathy of prematurity diagnosis *J. Intellig. Fuzzy Systems* **38** 6269–78

[48] Oloumi F, Rangayyan R M and Ells A L 2016 Computer-aided diagnosis of retinopathy in retinal fundus images of preterm infants via quantification of vascular tortuosity *J. Med. Imaging* **3** 044505

[49] Wallace D K, Jomier J, Aylward S R and Landers M B 2003 Computer-automated quantification of Plus disease in retinopathy of prematurity *J. Am. Assoc. Pediatr. Ophthalmol. Strabismus* **7** 126–30

[50] Lotmar W, Freiburghaus A and Bracher D 1979 Measurement of vessel tortuosity on fundus photographs *Graefe's Arch. Clin. Exp. Ophthalmol.* **211** 49–57

[51] Makkapati V V and Ravi V V C 2015 Computation of tortuosity of two dimensional vessels *8th Int. Conf. on Advances in Pattern Recognition (ICAPR)* (Piscataway, NJ: IEEE) pp 1–4

[52] Sivakumar R, Veena V and John R 2017 A curvature based approach for the automated screening of retinopathy of prematurity in preterm infants *13th Int. Conf. on Signal-Image Technology & Internet-Based Systems (SITIS)* (Piscataway, NJ: IEEE) pp 503–8

[53] Sivakumar R, Eldho M, Jiji C, Vinekar A and John R 2016 Diagnosis of Plus diseases for the automated screening of retinopathy of prematurity in preterm infants *11th Int. Conf. on Industrial and Information Systems (ICIIS)* (Piscataway, NJ: IEEE) pp 408–13

[54] Sivakumar R, Eldho M, Jiji C, Vinekar A and John R 2016 Computer aided screening of retinopathy of prematurity–a multiscale Gabor filter approach *6th Int. Symp. on Embedded Computing and System Design (ISED)* (Piscataway, NJ: IEEE) pp 259–26

IOP Publishing

Photo Acoustic and Optical Coherence Tomography Imaging, Volume 2

Fundus imaging for the retina

Ayman El-Baz and Jasjit S Suri

Chapter 7

Retinal disease management using fundus autofluorescence images

Shanmukh Reddy Manne, Kiran Kumar Vupparaboina, Gowtham Chowdary Gudapati, Ram Anudeep Peddoju, Chandra Prakash Konkimalla, Abhilash Goud, Sarforaz Bin Bashar, Jay Chhablani and Soumya Jana

7.1 Introduction

Medical imaging plays a crucial role in managing ophthalmic diseases. Among current modalities [1], fundus autofluorescence (FAF) imaging presents a relatively new technology that exploits the innate fluorescence of the retinal pigment epithelium (RPE), and has proven attractive as it does not require the injection of an artificial dye [2]. Specifically, the retina is illuminated with blue light which causes the naturally occurring pigment lipofuscin to fluoresce [3]. Consequently, the spatial distribution of intensity in the captured gray-scale FAF image indicates the distribution of lipofuscin. By inspecting such a distribution the ophthalmologist is able to visualize the status of the retina (particularly the RPE), and hence diagnose/monitor various ophthalmic diseases. More precisely, for disease diagnosis the physician generally hypothesizes a disease while inspecting an FAF image, and subsequently confirms it on the basis of quantitative indicators. For follow-up subjects, such indicators are used to monitor the disease course. In either scenario, the associated clinical assessment provides the basis for deciding further treatment, if required.

To facilitate such clinical assessment, significant effort has been directed at developing analysis tools specific to diseases, such as age-related macular degeneration (AMD) and Stargardt disease (STGD), an inherited ailment characterized by macular degeneration [4]. Such tools are based on disease-specific quantitative indicators, whose discovery has also remained an important focus. Indeed, indicators based on sectoral statistics collected from multi-segment grids centered at the fovea have been reported for various diseases including STGD [5], AMD [6, 7],

doi:10.1088/978-0-7503-2056-6ch7

central serous chorioretinopathy (CSCR) [8], and retinitis pigmentosa (RP) [9]. Further, owing to its high prevalence, special attention has been devoted to macular degeneration, studying variations in FAF phenotypes with disease progression [10], algorithmically quantifying associated geographic atrophy (GA) [11], characterizing such GA [12, 13], and investigating factors influencing the progression of such GA [14]. Quantification related to RP has also received significant attention [15].

However, advanced clinical assessment tools alone may not always ensure improved clinical outcomes. To appreciate this, consider subjects from economically disadvantaged backgrounds and/or remote regions, who face considerable economic/physical barriers in consulting trained ophthalmologists. In those circumstances, subjects tend to seek medical attention only after being reasonably certain that an adverse condition exists. Specifically, as shown in figure 7.1, such subjects seek a locally accessible economical screening service. If a diseased condition is indicated in the process, they become more likely to undertake the necessary yet arduous steps to obtain diagnosis, treatment, and treatment response monitoring, as needed. In this context, accurate screening assumes a crucial role in ensuring the desired clinical outcome. If many diseased subjects are missed, they undesirably go without treatment. In the other extreme, if the screening service is overly conservative and refers many normal subjects to further clinical assessment, it unnecessarily burdens a significant proportion of the subjects and keeps trained clinicians occupied with medically irrelevant cases, and eventually loses credibility. Thus the principal challenge here lies in ensuring a high level of screening accuracy without involving ophthalmologists. In response, we propose an accurate yet low-cost screening service which, by leveraging machine learning tools, requires only limited inputs from personnel with relatively little training.

Here, one important issue arises in choosing an appropriate imaging modality. As the development of such a screening tool has received rather limited attention in the context of eyecare, existing tools do not provide reliable guidance in making a choice. One recently suggested screening tool is based on fundus photography (FP) [16]. However, in this chapter we make use of FAF imaging in view of various extensive investigations that compared GA assessment based on FP images and that based on FAF images, and indicated higher consistency and reproducibility in the case of the

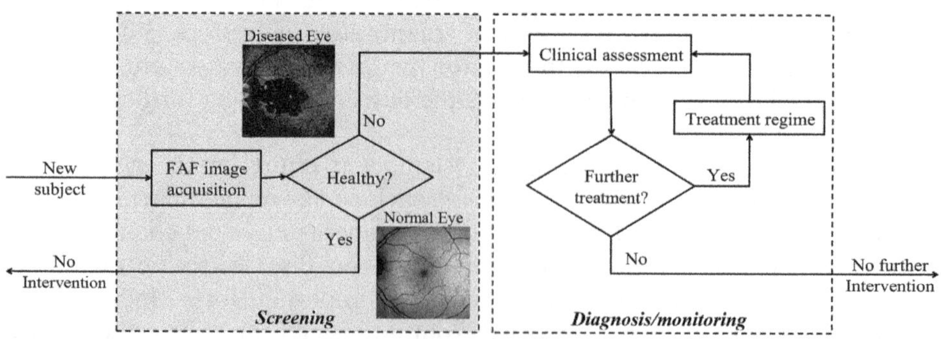

Figure 7.1. An end-to-end framework for a screening and diagnosis/treatment monitoring system.

latter modality [17–19]. More generally, FAF imaging has been found to facilitate the detection of abnormalities beyond those detected based on funduscopic exams, fluorescein angiography, or optical coherence tomography, and can be used to elucidate disease pathogenesis, form genotype–phenotype correlations, and diagnose and monitor disease [20]. Further, FAF imaging has been reported to lead to better understanding in disease-specific circumstances, including choroidal neo-vascularization (CNV) in AMD [21]. In short, FAF imaging has been understood to provide more reliable quantification compared to FP, but utilized principally in the context of diagnosis. In the present chapter, we propose FAF imaging as the basis for the screening of diseased eyes, in order to take advantage of its superiority. In addition, the aforesaid FP-based tool performed empirical mode decomposition, estimated the entropy and energy of each decomposed image, and used those as features to distinguish between FP images of healthy and diseased eyes [16]. However, such features are global, and do not capture local information about spatial variation, which, as alluded to earlier, disease-specific studies have found clinically relevant. Accordingly, for the proposed screening tool based on FAF imaging, we adopt a different set of clinically inspired features. In this context we hypothesize that the previously mentioned sectoral statistics of FAF images, based on which indicators for specific pathologies have already been suggested [6–9], also possesses general information about diseased conditions, and hence remain relevant even for screening. Specifically, given an FAF image, we propose to center an early treatment of diabetic retinopathy study (ETDRS) grid [22], a well-accepted tool in ophthalmic studies, at the fovea, and compute the mean and standard deviation of pixel intensity within each sector [6]. Collecting such sectoral statistics in a feature vector, a classifier is then trained to distinguish FAF images of healthy eyes from diseased eyes in such feature space.

Next, we turn to choosing a classifier for prediagnosis screening. Here, we also keep in mind that a patient often requires periodic monitoring of disease progression, post-diagnosis [10]. Accordingly, to keep the system cost low, we seek such a technological platform for the desired screening classifier that also potentially supports an efficient system for monitoring disease progression. In this regard, a recent work on retinal health screening adopted features based on empirical mode decomposition, and employed a support vector machine (SVM) classifier [16]. Such a classifier in general attempts to learn a decision boundary between diseased and healthy classes in the feature space from class labels [23] and is potentially attractive even for disease progress monitoring. Specifically, upon diagnosing a specific disease, one could note whether a patient's features gradually move further from (or closer to) the decision boundary over several visits, and possibly infer improvement in (or worsening of) the disease condition. In other words, an SVM classifier could supply the desired common platform alluded to earlier, provided that it achieves suitably high screening accuracy. Accordingly, the main focus of this chapter lies in establishing the suitability of SVM for ophthalmic disease screening using the proposed sectoral features.

At this point it is worthwhile to note that SVM classifiers can be implemented using different transformation kernels, the linear function and radial basis function (RBF) being the most ubiquitous ones [24]. Accordingly, we perform a comparative study between linear SVM and RBF-SVM in terms of screening performance, and report satisfactory performance of both the classifiers. However, we recommend the use of the latter, which exhibits a slight empirical superiority. In this connection, note that the aforesaid recent screening method also used RBF-SVM (albeit in association with a different set of features) but made no comparison with linear SVM [16]. Finally, to ensure improved fairness in evaluating the classifier performance, we adopt Monte Carlo cross validation (MCCV) rather than the more ubiquitous *K*-fold cross validation.

The rest of the chapter is organized as follows. FAF image acquisition, ETDRS grid placement and the computation of statistics, SVM classifiers, and performance evaluation criteria are described in section 7.2. In section 7.3 the experimental results on screening performance and statistical analysis are presented. Additional observations related to the performance of other classifiers and possible monitoring of disease progression using SVM are made in sections 7.4 and 7.5. The future course, covering potential extensions and practical implementation, is outlined in section 7.6.

7.2 Materials and methods

In this section we propose and describe a screening methodology, outlined in figure 7.2, where the ETDRS grid is placed on the input FAF image, and an SVM classier attempts to detect the occurrence of disease based on sectorial statistics. First, we detail the method of acquisition of FAF images.

7.2.1 Fundus autoflourescence image acquisition

This is a retrospective study conducted at the tertiary eye centers of L V Prasad Eye Institute (LVPEI) in South India. A total of 140 subjects underwent fundus autoflourescence examination where 61 were healthy and the rest were diagnosed to have ophthalmic diseases. Specifically, of the 79 subjects, 21 were diagnosed with CSCR, 14 with choroidal neovascular membranes (CNVM), and 44 with STGD. The procedures performed in this study conformed to the tenets of the Declaration of Helsinki. Respective Institutional Review Boards approved the study protocol.

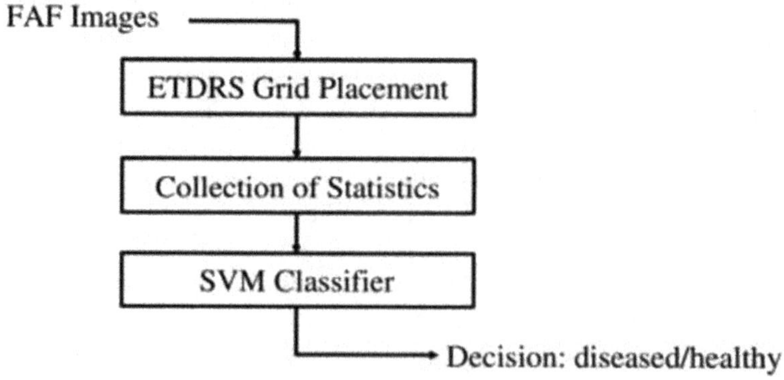

Figure 7.2. Schematic diagram of the proposed screening method.

Written informed consent was obtained from all the study participants. The Heidelberg Retina Angiograph 2 (HRA2; Heidelberg Engineering, Heidelberg, Germany) was used to acquire the FAF images. The standard protocol for image acquisition involved excitation at 488 nm with an optically pumped solid-state laser and emission detected above 500 nm with a barrier filter. Typically, between 9 and 15 images were averaged to acquire a mean image. Either the automatic real-time mode or the mean image mode was used for all FAF image acquisition. This protocol allowed for an amplified signal with reduced noise for each image.

7.2.2 ETDRS-based statistics as the feature set

In the current work, the ETDRS grid was used for computing features from each FAF image with the goal of classifying it as either 'healthy' or 'diseased'. The ETDRS grid, positioned by an optometrist (rather than an ophthalmologist) at the center of the fovea as shown in figure 7.3, has nine segments. The inner most ring, known as central subfield (CSF), is surrounded by the pericentral ring consisting of the temporal inner macula (TIM), inferior inner macula (IIM), nasal inner macula (NIM) and superior inner macular (SIM) regions. Further, the peripheral ring in the ETDRS grid consists of the temporal outer macula (TOM), inferior outer macula (IOM), nasal outer macula (NOM), and superior outer macular (SOM) regions. In each of the nine segments, the mean pixel intensity and the corresponding standard deviation were computed. Such statistics from all nine segments formed a feature

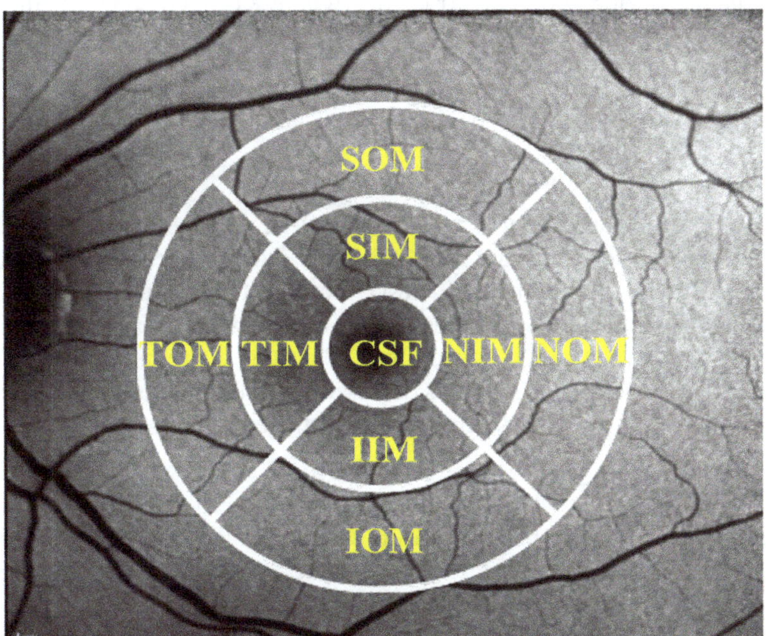

Figure 7.3. FAF image with the ETDRS grid.

vector of length 18 representing each FAF image. These features were used to distinguish between diseased and healthy eyes.

7.2.3 Support vector machine (SVM) classifier

The support vector machine (SVM) classifier provides a supervised learning model [23] which, upon training, obtains a hyperplanar decision boundary maximally separating two classes. More generally, to handle data that are nonlinearly separable, SVM can incorporate a suitable kernel that transforms the original feature space into a higher-dimensional space, where the transformed features become linearly separable (see figure 7.4 for an explanatory illustration). Formally, the decision hyperplane $w^T\varphi(x) + b = 0$ (where $\varphi(x)$ denotes a point vector in the transformed feature space, φ the kernel function, w the weight vector, and b the bias) is chosen to maximize overall separation. One equivalently maximizes

$$\sum_{i=1}^{n}\alpha_i - 0.5\sum_{i=1}^{n}\sum_{j=1}^{n}\alpha_i\alpha_j y_i y_j K(x_i, x_j).$$

over $\alpha_i \geqslant 0$, $i = 1, 2, \ldots, n$, subjects to $\sum_{i=1}^{n}\alpha_i y_i = 0$. Here, for each data index i, x_i and y_i, respectively, indicate the feature vector and the corresponding class label (+1 for diseased; −1 for healthy), and K denotes the kernel function, n the size of the dataset, and α_i the Lagrange multiplier. We compare the classification performance of two variants of the SVM, one with a linear kernel (where $K(x_i, x_j) = x^T x_j$), and the other with a radial basis function (where $K(x_i, x_j) = \exp\frac{|x_i - x_j|^2}{2\sigma^2}$), an ubiquitous nonlinear kernel.

7.2.4 Performance evaluation

To evaluate the classification performance, the SVM parameters, namely the weight vector w and bias b, are optimized on a subset of data (the training subset) and are validated against the complementary (test) subset. More accurately, the dataset is partitioned into training and test subsets such that (i) the ratio of their sizes, called the training-to-test (split) ratio, approximately equals a pre-assigned fraction, and (ii) the proportion of healthy (as well as diseased) images represented in those subsets also approximately equals the same fraction. In general, the performance of the

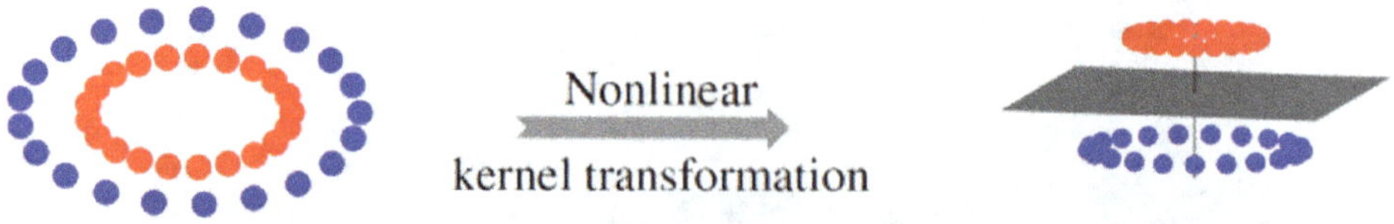

Figure 7.4. SVM with a nonlinear kernel. Data points belonging to each class are indicated by unique color; potential linear separability upon nonlinear kernel transformation.

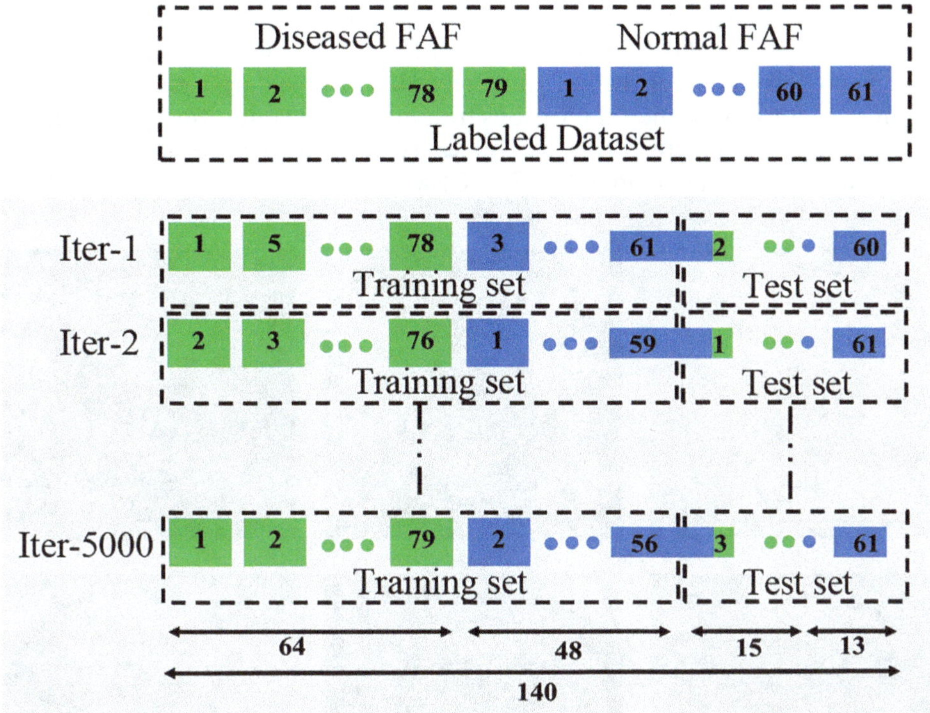

Figure 7.5. Illustration of the MCCV method.

classifier depends on the subjective choice of the partition. To avoid such subjectivity in performance analysis, one customarily uses Monte Carlo cross validation (MCCV) [25] where, as shown in figure 7.5, the dataset is randomly partitioned a large number (5000) of times (iterations) keeping the split ratio as well as the aforementioned proportions constant. For each iteration, the SVM parameters are optimized over the training subset, and the mean training as well as test accuracy and corresponding standard deviation are recorded. Noting that the training accuracy indicates the classification performance for seen data, while the test accuracy indicates that for unseen data, classifiers with high average test accuracy assume practical significance. Further, low standard deviation, indicating low performance variability over random partitions and hence signifying robustness, is desirable. Additionally, we also compute the average confusion matrix over 5000 iterations, which provides the class-conditional detection probability for each of the healthy and diseased classes. Finally, we train and evaluate classifiers for different split ratios.

7.3 Experimental results

As described earlier, placing an ETDRS grid on each FAF image, we computed the mean and standard deviation of pixel values corresponding to each of the nine sectors (figure 7.3) and thus formed a feature vector of length 18. In figure 7.6 such sectoral features are illustrated for example FAF images of both healthy and diseased eyes. As mentioned earlier, two variants of SVM classifiers, a linear SVM and an SVM with an RBF kernel (RBF-SVM), were considered. In each case, the training and test accuracy values were recorded over a large number (5000) of random partitions for each of the various training-to-test split ratios chosen between 10:90 and 90:10, and the average values (with standard deviation values in parentheses) are tabulated in table 7.1. Clearly, in the case of linear SVM, the

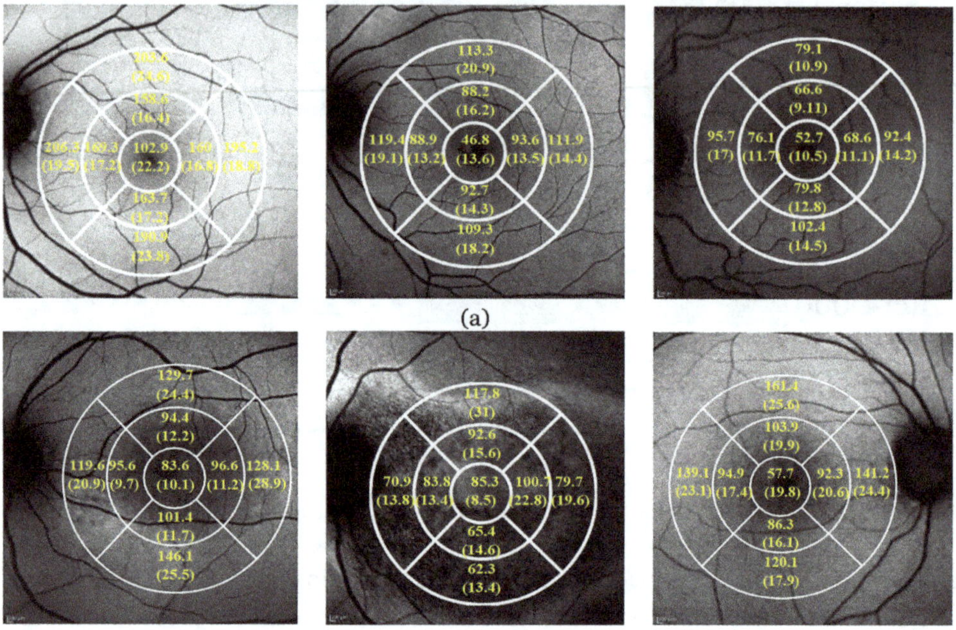

Figure 7.6. ETDRS grid placement on and corresponding sectorial statistics (mean with standard deviation in parenthesis) for representative images of (a) healthy eyes and (b) diseased eyes.

Table 7.1. Performance of rival SVM classifiers for different splits (superior value between linear SVM and RBF-SVM boldfaced).

	Linear SVM			RBF-SVM		Gain (%)	
Split ratio	Train. acc. a (b)	Test acc. c (d)	SF	Train. acc. e (f)	Test acc. g (h)	Train. acc. $e-a$ ($^{b-f}$)	Test acc. $^{g-c}$($d-h$)
10:90	91.27 (7.01)	75.03 (7.05)	2.85	**99.38 (2.13)**	**78.97 (5.30)**	8.16 (69.61)	4.99 (24.82)
20:80	91.51 (4.82)	82.26 (4.78)	2.80	**98.81 (2.09)**	**84.32 (3.96)**	7.39 (56.65)	2.44 (17.15)
30:70	92.04 (3.66)	85.56 (3.81)	3.20	**97.88 (2.17)**	**86.66 (3.48)**	5.97 (40.71)	1.27 (8.66)
40:60	92.26 (2.91)	87.31 (3.46)	2.70	**98.86 (1.38)**	**88.11 (3.35)**	6.67 (52.27)	0.91 (3.18)
50:50	92.24 (2.44)	88.40 **(3.43)**	3.00	**98.86 (1.38)**	**88.89 (3.43)**	6.69 (43.44)	0.55 (0.00)
60:40	92.27 (2.01)	89.02 (3.68)	2.90	**98.46 (1.20)**	**89.58 (3.47)**	6.28 (40.29)	0.62 (5.71)
70:30	92.31 (1.64)	89.53 **(4.14)**	2.80	**98.62 (0.99)**	**90.10** (4.16)	6.39 (39.63)	0.63 (−0.48)
80:20	92.26 (1.30)	89.88 **(5.12)**	2.75	**98.65 (0.90)**	**90.55** (5.24)	6.47 (30.77)	0.75 (−2.34)
90:10	92.25 (1.01)	89.60 **(7.16)**	2.65	**98.86 (0.71)**	**90.83** (7.21)	6.68 (29.70)	1.35 (−0.70)

average training as well as test accuracy level tends to increase with an increasing training-to-test ratio but for a few exceptions. In other words, as expected, the said classifier tends to learn better with increased availability of training data. Turning to the RBF-SVM classifier, for each split ratio we chose the scale factor (SF) that maximizes test accuracy. The resulting average test accuracy, which still largely follows the aforementioned increasing trend, slightly improves over that obtained using the linear SVM for each split ratio. This indicates a moderate degree of nonlinearity inherent in the underlying problem. Despite only a slight gain in test accuracy, the gain in average training accuracy is significantly higher, indicating that the RBF kernel is closely modeling certain nonlinear aspects of the training data that do not generalize well. In fact, a similar conclusion can also be drawn by considering standard deviation values as follows. Note that the standard deviation in training accuracy decreases with increasing split ratio in both cases of linear SVM and RBF-SVM; however, the values are significantly lower for the latter classifier, indicating better modeling. Yet, turning to test accuracy, the standard deviation is lower in the case of RBF-SVM for lower values of split ratio and in the case of linear SVM for higher values of split ratio. This phenomenon does not indicate improved generalization. Further, the standard deviation in either case first decreases and then increases with increasing split ratio. Yet a subtle difference can be discerned. While the test accuracy is most reliable (i.e. with the lowest standard deviation) for an evenly split training-to-test ratio in the case of linear SVM (also observed elsewhere [26]), the highest reliability is observed at the skewed split ratio of 40:60 in the case of RBF-SVM.

Having reported performance variation over the entire range of split ratios, we next delve deeper into the case of the 80:20 split, a general recommendation for various practical applications [27]. First, we studied the performance variation with varying SF. In particular, the SF was varied between 1 and 5, and corresponding training and test accuracy levels over 5000 MCCV iterations were recorded. Notably, as seen in the figure 7.7, the training accuracy decreases with an increase in SF, whereas the test accuracy increases with an increase in SF up to a maximum value of 2.75. For larger values of SF, the test accuracy decreases so slowly that it may essentially be taken as unchanged. Thus we take RBF-SVM with SF = 2.75 as the classifier competing with the linear SVM. In particular, we next study the class-conditional performance of these rival classifiers at hand. The practical significance of this study arises from the fact that mistakenly classifying a diseased eye as normal leads to the severe consequence of denial of treatment to a *bona fide* patient, whereas declaring a normal eye as diseased leads to the comparatively less severe outcome of wasted time and resources. Referring to table 7.2, the RBF-SVM records an average probability of 91.61% in correctly detecting the disease class, whereas it is only 89.47% using the linear SVM, with respective standard deviations of 7.09% and 7.33%. In contrast, conditioned on the healthy class, the healthy class was correctly detected with an average probability of 92.25% by the linear SVM and 88.59% by RBF-SVM, with respective standard deviations of 7.11% and 8.82%. Thus, given the healthy class, linear SVM slightly outperforms RBF-SVM in terms of both accuracy and reliability. Note that the advantage of RBF-SVM over linear SVM given the diseased class is comparable to that of linear SVM over RBF-SVM given the healthy class. However,

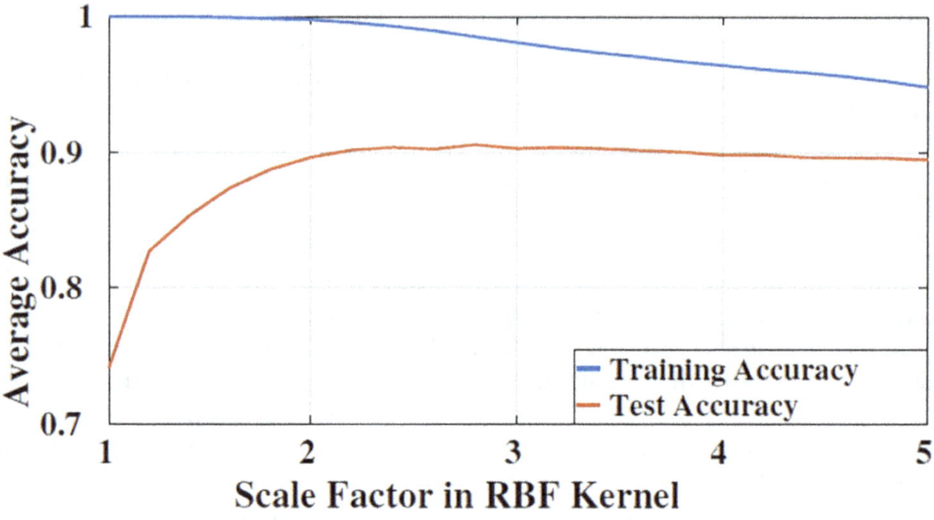

Figure 7.7. Average training and test accuracy of the RBF-SVM classifier over MCCV with variation in scale factor (80:20 split ratio).

Table 7.2. Test confusion matrices (%) of linear SVM and RBF-SVM for a split ratio 80:20 for the original screening dataset.

	Actual/Predicted	Diseased	Healthy
Linear SVM	Diseased	89.47 (7.33)	7.75 (7.11)
	Healthy	10.53 (7.33)	92.25 (7.11)
RBF-SVM	Diseased	91.61 (7.09)	11.41 (8.82)
	Healthy	8.39 (7.09)	88.59 (8.82)

since making an error incurs higher practical cost given the diseased class than that given the healthy class, one should prefer RBF-SVM over linear SVM as a screening tool. In summary, based on our experiments, we conclude that SVM classifiers of FAF images, attaining close to 90% accuracy, provide attractive screening

Finally, it is worthwhile to demonstrate that the screening problem poses harder challenges compared to detection of a specific disease. Recall that our screening dataset consists of FAF images of healthy eyes as well as eyes affected by the highly prevalent diseases STGD, CNVM, and CSCR, between which we did not distinguish so far. At this point, consider the disease detection problem instead, where one checks whether a particular disease (say, Stargardt) is present. Specifically, we now decide between healthy versus Stargardt-affected eyes, and note for RBF-SVM the marked improvement in the corresponding class-conditional accuracy levels (99.35% and 99.02%, respectively) given in table 7.3 over those (91.61% and 88.59%, respectively; see table 7.2) in the case of disease-agnostic screening, indicating the relative difficulty posed by the screening problem.

Table 7.3. Test confusion matrix (%) of the RBF-SVM for a split ratio of 80:20 for the reduced dataset (diseased class represented by STGD only) tools for ophthalmic diseases. Among those, the RBF-SVM appears to provide a slight advantage over the linear SVM in terms of both overall accuracy levels and practical class-conditional costs.

Actual/Predicted	Diseased	Healthy
Diseased	99.35 (2.65)	0.98 (2.61)
Healthy	0.65 (2.65)	99.02 (2.61)

7.4 Additional performance comparison

So far our focus has remained on the development of an accurate SVM-based automated screening tool that detects the possible presence of ophthalmic disease. In this context one would wonder how the proposed SVM-based classifier compares in terms of performance with other well-known alternatives that have proven efficacious in related contexts. Accordingly, we considered partial least squares trained linear discriminant analysis (PLS-LDA1) [28] (effective in diagnostic quality assessment of fundus photographs [29]), extreme gradient boosting of decision trees (XGBoost) [30] (effective in classifying retinal cysts [31]), and random forest (RF2) classifiers [32] (effective in detecting diabetes [33], diabetic retinopathy, and age-relate macular degeneration based on fundus images [34]). For the present screening problem, we plotted in figure 7.8 the average test accuracy obtained, respectively, by the SVM variants and other classifiers at hand against variation in training-to-test split ratio. We observe that the proposed SVM-RBF classifier outperforms the rivals across the split ratios. In fact, both SVM classifiers generally prove superior to the rest of the classifiers under consideration. As an aside, PLS-LDA, while inferior to the SVM classifiers, performs superior to RF as well as XGBoost at the 80:20 split. Here, the hyperparameters were tuned (details omitted for the sake of brevity and continuity) for each of the PLS-LDA, XGBoost, and RF classifiers as well, so as to ensure fairness of comparison with the tuned SVM classifiers.

7.5 Possible progress monitoring

As alluded to earlier, we envisage an SVM-based progress monitoring system, where FAF images of an eye are taken over multiple patient visits. In this setting, one would consider the diagnosed disease and compute the time course of signed (positive for normal; negative for diseased class) distances of the corresponding feature vectors from the disease-specific decision boundary based on successive FAF images. For a correctly diagnosed subject, the trajectory should begin within the diseased class. Further, a trajectory heading towards the decision boundary would indicate gradual improvement, while that moving further away would indicate a deteriorating condition. Here, while making a feasibility study of the envisaged system for disease progress monitoring in terms of class separation, we restrict ourselves to only RBF-SVM classifiers.

To this end, we first revisit the proposed RBF-SVM screening system, and set the corresponding class separation as a reference. In particular, we obtain for each fixed split ratio the optimum decision boundary corresponding to each of the 5000 random partitions mentioned in section 7.3. Further, fixing each such boundary, we compute the signed (Euclidean) distance of each feature vector (corresponding to an FAF image from the training as well as the test subset) from that boundary. Next, corresponding to each FAF image (data point), we obtain a collection of 5000 such distances, and calculate the mean and the standard deviation. In figure 7.9(a) we

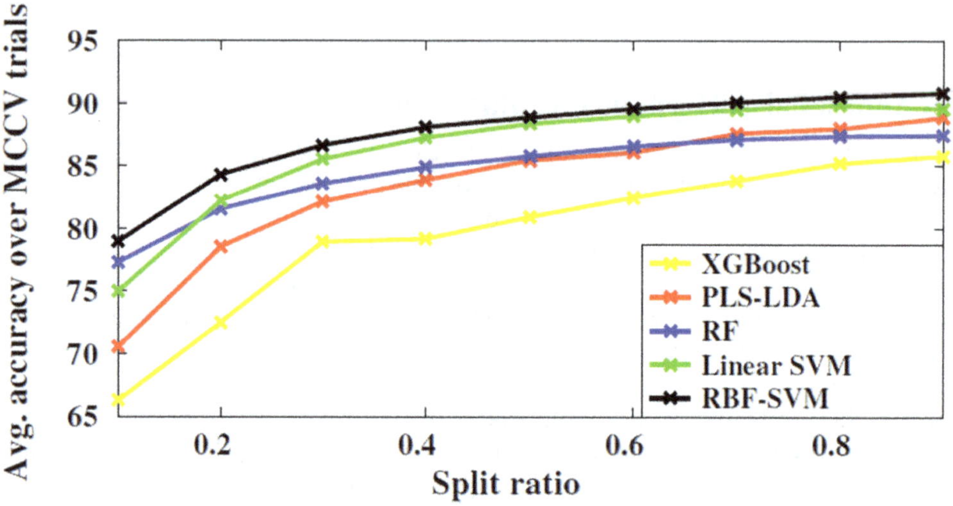

Figure 7.8. Superiority of RBF-SVM over other classifiers across a varying split ratio.

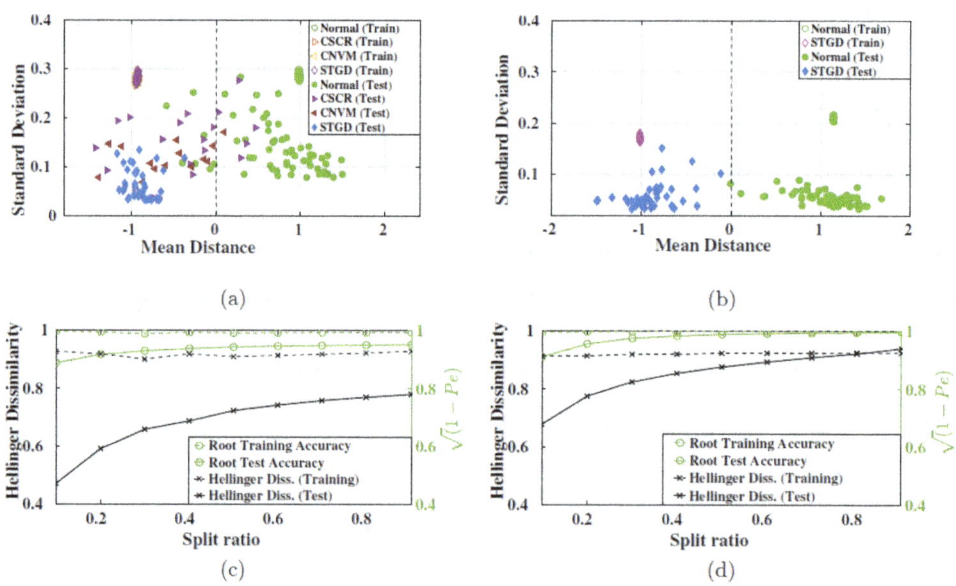

Figure 7.9. Signed distances from the decision boundary over all iterations of MCCV for the RBF-SVM on (a) the original dataset and (b) the reduced dataset; variation of the Hellinger dissimilarity and root accuracy against the split ratio on (c) the original dataset and (d) the reduced dataset.

consider the fixed split ratio 80:20, and obtained a scatter plot of the aforementioned quantities by sweeping the said data point through the entire screening dataset at hand. Here, recall again that our screening dataset consists of FAF images of healthy eyes as well as eyes affected by STGD, CNVM, and CSCR. In the above plot, we distinguish between different diseases only as a visualization aid (without making any disease-specific analysis).

Post-diagnosis, the disease affecting each subject would be known (ignoring misdiagnosis) and one potentially needs to monitor subsequent progress of that disease. In this context, the following analysis is presented for the case when the diagnosed disease is STGD in view of its relatively high prevalence. In this regard, we next considered the subset of our dataset that consists of images of only STGD-affected and healthy eyes, and ignored the rest. On this reduced dataset, as earlier, we considered the 80:20 split ratio, and again obtained optimum decision boundaries between diseased (representing only STGD now) and healthy classes using the RBF-SVM classifier over 5000 random partitions, and subsequently computed the desired distances from respective decision boundaries. A scatter plot of the image (data point)-wise mean and standard deviation of those distances are analogously furnished in figure 7.9(b). Comparing figure 7.9(b) with figure 7.9(a), the classes are observed to be more clearly separated when only STGD-affected eyes represent the diseased class. More generally, we expect class separation for any disease-specific reduced dataset to be larger than that for the original screening database.

At this point, we desired to study the above phenomenon more generally across all split ratios. However, as inspection of a series of scatter plots could be subjective and tedious, we now resorted to quantitative comparison of class separation. In particular, we used the Hellinger dissimilarity (HD) between probability distributions p and q [35], defined by

$$H = \left(1 - \sum_{x} p^{\frac{1}{2}}(x) q^{\frac{1}{2}}(x)\right)^{\frac{1}{2}},$$

as a measure of separation between the healthy and diseased classes, by taking p and q to be the respective distributions of distances corresponding to the aforesaid classes, and denoting the distance variable by x.

Specifically, for each split ratio, we estimated the probability distribution p (respectively q) of the aforementioned distances corresponding to the healthy (respectively diseased) class over 5000 partitions of training as well as test sets, as customary. Using these distributions we obtained the HD measure, which is plotted against the split ratio for both the training and test sets. In particular, presented in figure 7.9(c), the aforesaid plots for the original dataset, representing STGD, CNVM, and CSCR, show an increasing HD with increasing split ratio for the test cases, and a more-or-less constant behavior with slight fluctuations in the case of training. Further, the HD for training remains greater than that for test throughout. Shown in figure 7.9(d), analogous plots for the reduced dataset representing STGD only, also reveal an almost constant HD for training, nearly matching that for the original dataset with even fewer fluctuations. As expected, the test HD for the reduced dataset, while still exhibiting an increasing behavior, is significantly greater than that for the original dataset for any given split ratio. As an interesting aside, the test HD now crosses over the training HD to become the larger quantity for split ratios slightly greater than 0.8.

Thus, in view of the significant separation between a specific disease (demonstrated for STGD) and the healthy class, the RBF-SVM potentially provides a satisfactory tool for monitoring of disease progress. Of course, here we provide only preliminary evidence, and it remains essential to undertake a confirmatory study involving multiple visits of several post-diagnosis patients in order to draw firm conclusions.

7.6 Discussion and future course

In this chapter we developed a screening technique that does not require the participation of ophthalmologists, and semi-automatically identifies the presence of disease based on FAF images with minimal assistance from optometrists. Specifically, we proposed an RBF-SVM classifier, and demonstrated its efficacy in terms of not only classification accuracy, but also other desirable properties. Such a technique holds vast potential in transforming eyecare in remote regions, particularly in developing countries such as India. Indeed, existing organizations, including the L V Prasad Eye Institute (LVPEI), which has a hierarchical organizational structure with primary centers serving far-flung areas, could be of particular help in propagating the proposed screening service. In turn, such a service, apart from screening remote subjects, can potentially make the said organizations more efficient, by referring only medically relevant subjects for diagnosis and/or treatment at secondary or tertiary centers. Further, in view of the excellent class separation exhibited by the proposed classifier, our screening service can potentially be integrated with post-diagnosis treatment response monitoring. In this regard, the distances so far have been measured from the partition-specific decision boundary, the main challenge lies in selecting a representative partition-independent boundary. However, noting low variability in MCCV performance, we believe that the choice of such a representative boundary should not be very difficult. Finally, while FAF imaging presents certain broad advantages over FP, the latter has benefits of its own. Accordingly, in the future, we plan to incorporate FP alongside FAF imaging to improve screening (as well as disease progress monitoring) performance.

Appendix A Theoretical connection

While development and demonstration of a screening tool remained our primary focus, we seek to point out an information-theoretic connection that may interest certain readers. To begin with, revisit figure 7.9(c) and (d), and consider the variation in the square root of classification accuracy levels in various cases, distinguishing between the training and test scenarios.

Comparing analogous cases, we observe that both the root classification accuracy and the HD measure exhibit somewhat similar variational behavior, although the former always remains greater than the latter.

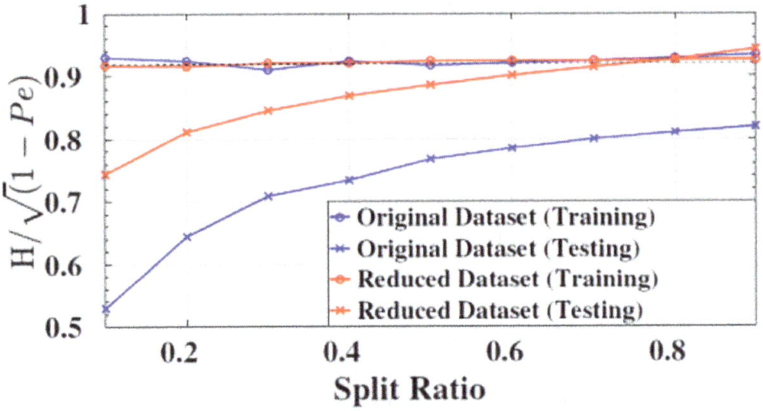

Figure A.1. Variation of fraction $H/\sqrt{1-P_e}$ against the split ratio on both the datasets.

This observation is in fact consistent with an information-theoretic principle. To see this, consider the well-known Chernoff bound

$$P_e < \sum_x p^{1-s}(x)q^s(x), \tag{A.1}$$

which holds for any exponent $0< s <1$ [36]. Taking $s = \frac{1}{2}$, using the definition of HD and reorganizing (A.1), we obtain

$$H < \sqrt{1-P_e}. \tag{A.2}$$

The bound (A.2) explains the observed dominance of root classification accuracy over the HD.

Further, the proposed RBF-SVM classifier attempts to maximize the class separation in the training subset for each split ratio, bringing H and $\sqrt{1-P_e}$ close. In fact, for training, we observe the fraction $H/\sqrt{1-P_e}$ to hover around a constant 0.92 across various split ratios, for both the original and the reduced datasets (see figure A.1). This indicates that the complexity of maximization of class separation which underlies training remains similar for competing datasets across split ratios. However, in the test scenarios such a fraction exhibits an increasing behavior against the split ratio, indicating that the gap between H and $\sqrt{1-P_e}$ in (A.2) narrows when a higher proportion of training data becomes available, as anticipated. Further, the said fraction for the original dataset is found to be smaller than that for the reduced dataset for every split ratio, reflecting the decrease in problem complexity of the latter disease-specific problem. In fact, for the reduced dataset, the test fraction crosses the training fraction for large split ratios.

Acknowledgments

The work was partly supported by Grant BT/PR16582/BID/7/667/2016, Department of Biotechnology (DBT), Ministry of Science and Technology, the Government of India. Shanmukh Reddy Manne thanks the Ministry of Electronics and Information Technology (MeitY), the Government of India, for a fellowship grant under the Visvesvaraya PhD Scheme for Electronics and IT.

References

[1] Yolcu U, Sahin O F and Gundogan F C 2014 Imaging in ophthalmology *Ophthalmology: Current Clinical and Research Updates* (London: IntechOpen)

[2] Nickla D L and Wallman J 2010 The multifunctional choroid *Progr. Retinal Eye Res* **29** 144–68

[3] Schmitz-Valckenberg S, Holz F G, Bird A C and Spaide R F 2008 Fundus autofluorescence imaging: review and perspectives *Retina* **28** 385–409

[4] Smith R T, Lee N, Chen J, Busuioc M and Laine A F 2008 Interactive image analysis in age-related macular degeneration (AMD) and Stargardt disease (STGD) *42nd Asilomar Conf. on Signals, Systems and Computers* (Piscataway, NJ: IEEE) pp 651–4

[5] Burke T R, Duncker T, Woods R L, Greenberg J P, Zernant J, Tsang S H, Smith R T, Allikmets R, Sparrow J R and Delori F C 2014 Quantitative fundus autofluorescence in recessive Stargardt disease *Investigative Ophthalmol. Vis. Sci.* **55** 2841–52

[6] Marsiglia M, Boddu S, Bearelly S, Xu L, Breaux B E, Freund K B, Yannuzzi L A and Smith R T 2013 Association between geographic atrophy progression and reticular pseudodrusen in eyes with dry age-related macular degeneration *Investigative Ophthalmol. Vis. Sci.* **54** 7362–9

[7] Schmitz-Valckenberg S, Fleckenstein M, Göbel A P, Sehmi K, Fitzke F W, Holz F G and Tufail A 2008 Evaluation of autofluorescence imaging with the scanning laser ophthalmoscope and the fundus camera in age-related geographic atrophy *Am. J. Ophthalmol.* **146** 183–92

[8] Zola M, Chatziralli I, Menon D, Schwartz R, Hykin P and Sivaprasad S 2018 Evolution of fundus autofluorescence patterns over time in patients with chronic central serous chorioretinopathy *Acta Ophthalmol.* **96** e835–9

[9] Jolly J K, Wagner S K, Moules J, Gekeler F, Webster A R, Downes S M and MacLaren R E 2016 A novel method for quantitative serial autofluorescence analysis in retinitis pigmentosa using image characteristics *Transl. Vis. Sci. Technol.* **5** 10

[10] Schmitz-Valckenberg S, Fleckenstein M, Scholl H P and Holz F G 2009 Fundus autofluorescence and progression of age-related macular degeneration *Surv. Ophthalmol.* **54** 96–117

[11] Schmitz-Valckenberg S, Brinkmann C K, Alten F, Herrmann P, Stratmann N K, Göbel A P, Fleckenstein M, Diller M, Jaffe G J and Holz F G 2011 Semiautomated image processing method for identification and quantification of geographic atrophy in age-related macular degeneration *Investigative Ophthalmol. Vis. Sci.* **52** 7640–6

[12] Fleckenstein M, Schmitz-Valckenberg S, Martens C, Kosanetzky S, Brinkmann C K, Hageman G S and Holz F G 2011 Fundus autofluorescence and spectral-domain optical coherence tomography characteristics in a rapidly progressing form of geographic atrophy *Investigative Ophthalmol. Vis. Sci.* **52** 3761–6

[13] Holz F G *et al* 2007 Progression of geographic atrophy and impact of fundus autofluorescence patterns in age-related macular degeneration *Am. J. Ophthalmol.* **143** 463–72

[14] Fleckenstein M, Mitchell P, Freund K B, Sadda S, Holz F G, Brittain C, Henry E C and Ferrara D 2018 The progression of geographic atrophy secondary to age-related macular degeneration *Ophthalmology* **125** 369–90

[15] Schuerch K, Woods R L, Lee W, Duncker T, Delori F C, Allikmets R, Tsang S H and Sparrow J R 2017 Quantifying fundus autofluorescence in patients with retinitis pigmentosa *Investigative Ophthalmol. Vis. Sci.* **58** 1843–55

[16] Acharya U R, Mookiah M R K, Koh J E, Tan J H, Bhandary S V, Rao A K, Fujita H, Hagiwara Y, Chua C K and Laude A 2016 Automated screening system for retinal health using bi-dimensional empirical mode decomposition and integrated index *Comput. Biol. Med.* **75** 54–62

[17] Bearelly S, Chen M, Kelly M and Cousins S 2007 Comparing fundus autofluorescence imaging with fundus photographs in patients with geographic atrophy related to age-related macular degeneration *Investigative Ophthalmol. Vis. Sci.* **48** 2163–3

[18] Khanifar A A, Lederer D E, Ghodasra J H, Stinnett S S, Lee J J, Cousins S W and Bearelly S 2012 Comparison of color fundus photographs and fundus autofluorescence images in measuring geographic atrophy area *Retina* **32** 1884

[19] Hadi M A *et al* 2013 Comparison between fundus autofluorescence images and color fundus photos in patients with late dry age-related macular degeneration *Egypt. Retina J.* **1** 31

[20] Yung M, Klufas M A and Sarraf D 2016 Clinical applications of fundus autofluorescence in retinal disease *Int. J. Retina Vitreous* **2** 12

[21] Dandekar S S, Jenkins S A, Peto T, Scholl H P, Sehmi K S, Fitzke F W, Bird A C and Webster A R 2005 Autofluorescence imaging of choroidal neovascularization due to age-related macular degeneration *Archiv. Ophthalmol.* **123** 1507–13

[22] Early Treatment Diabetic Retinopathy Study Research Group *et al* 1991 Early photocoagulation for diabetic retinopathy: ETDRS report number 9 *Ophthalmology* **98** 766–85

[23] Cortes C and Vapnik V 1995 Support-vector networks *Mach. Learn.* **20** 273–97

[24] Chang C-C and Lin C-J 2011 LIBSVM: a library for support vector machines *ACM Trans. Intelligent Syst. Technol.* **2** 27

[25] Xu Q-S and Liang Y-Z 2001 Monte Carlo cross validation *Chemometr. Intell. Lab. Syst.* **56** 1–11

[26] Tripathy S, Reddy M S, Vanjari S R K, Jana S and Singh S G 2019 A step towards miniaturized milk adulteration detection system: smartphone-based accurate pH sensing using electrospun halochromic nanofibers *Food Anal. Methods* **12** 612–24

[27] Macek K 2008 Pareto principle in datamining: an above-average fencing algorithm *Acta Polytechnica* **48** 6

[28] Qiao Z, Zhou L and Huang J Z 2008 Effective linear discriminant analysis for high dimensional, low sample size data *Proc. World Congress on Engineering Lecture Notes in Engineering and Computer Science* vol 2 (Hong Kong: Newswood) pp 2–4

[29] Dev C, Sharang M, Manne S R, Goud A, Bashar S B, Richhariya A, Chhablani J, Vupparaboina K K and Jana S 2019 Diagnostic quality assessment of ocular fundus photographs: efficacy of structure-preserving ScatNet features *41st Annual Int. Conf. of the IEEE Engineering in Medicine and Biology Society (EMBC)* (Piscataway, NJ: IEEE) pp 2091–4

[30] Chen T and Guestrin C 2016 XGBoost: a scalable tree boosting system *Proc. 22nd ACM SIGKDD Int. Conf. on Knowledge Discovery and Data Mining* (New York: ACM) pp 785–94

[31] Teja R V, Manne S R, Goud A, Rasheed M A, Dansingani K K, Chhablani J, Vupparaboina K K and Jana S 2019 Classification and quantification of retinal cysts in OCT B-scans: efficacy of machine learning methods *41st Annual Int. Conf. of the IEEE Engineering in Medicine and Biology Society (EMBC)* (Piscataway, NJ: IEEE) pp 48–51

[32] Breiman L 2001 Random forests *Mach. Learn.* **45** 5–32

[33] Samant P and Agarwal R 2018 Machine learning techniques for medical diagnosis of diabetes using iris images *Comput. Methods Programs Biomed.* **157** 121–8

[34] Morales S, Engan K, Naranjo V and Colomer A 2015 Detection of diabetic retinopathy and age-related macular degeneration from fundus images through local binary patterns and random forests *IEEE Int. Conf. on Image Processing (ICIP)* (Piscataway, NJ: IEEE) pp 4838–42

[35] Deza M-M and Deza E 2006 *Dictionary of Distances* (Amsterdam: Elsevier)

[36] Poor H V 2013 *An Introduction to Signal Detection and Estimation* (Berlin: Springer)

IOP Publishing

Photo Acoustic and Optical Coherence Tomography Imaging, Volume 2
Fundus imaging for the retina
Ayman El-Baz and Jasjit S Suri

Chapter 8

A review of mainstream ophthalmic AI algorithms: advances, limitations, and challenges

Lie Ju, Yicheng Wu, Lin Wang, Wei Feng, Di Xu, Xin Wang, Xin Zhao and Zongyuan Ge

8.1 Brief introduction

A fundus image is a projection of the fundus captured by a monocular camera on a 2D plane. The main structures that can be visualized in a fundus photograph are the central and peripheral retina, optic disc, and macula (figure 8.1).

Recently, deep learning-based methods have shown promising accuracy for automated retinal disease screening. There are several papers [1–5] that have performed a comprehensive review of recent advances in computer-assisted diagnosis (CAD) of retinal diseases. For instance, Li *et al* [5] provided a high-quality review that focused on the successful application of deep learning methods in fundus

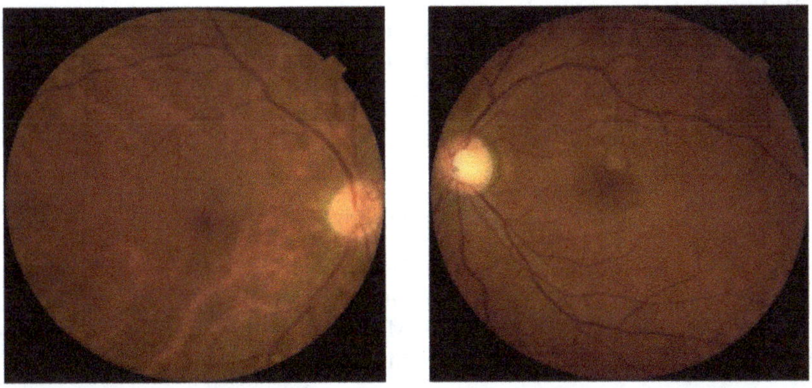

Figure 8.1. Normal fundus photographs of the right eye (left panel) and left eye (right panel).

doi:10.1088/978-0-7503-2056-6ch8

images from January 2016 to August 2020 with 143 papers collected. The literature are organized by the target tasks and disease types in detail.

Contributions: In this work, different from the above reviews, we only focus on those mainstream directions that have relatively substantial research and show the opportunity to be applied in clinical or practical scenarios. We first briefly summarize the accessible public datasets, and some related works are presented. Finally, we will discuss the limitations of existing works from the perspective of both research and practical applications. Then, we also give some potential solutions and future research directions. We hope this review can provide inspiring ideas for researchers in the community.

8.2 Classification/detection

8.2.1 Diabetic retinopathy classification/grading

Diabetic retinopathy (DR) is one kind of common retinal disease that leads to a serious impact on vision. According to the International Clinical Diabetic Retinopathy Scale (ICDRS), DR can be graded into five levels, which are normal, non-proliferative DR-I (NPDRI), non-proliferative DR-II (NPDRII), non-proliferative DR-III (NPDRIII), and proliferative DR (PDR). Detailed descriptions and diagnostic criteria are summarized in table 8.1.

8.2.1.1 Datasets

Diabetic retinopathy has long been recognized and there are many public datasets in the community. In this review we only take those with more than 500 images into consideration and we provide statistics for the datasets in table 8.2. Also, it should be noted that some of thoese datasets also have a lesion-level annotation, such as hemorrhages and exudates, which occur widely in DR, and we provide this part of the discussion in section 8.3.1. In addition to the public datasets mentioned above, most clinical papers prefer to use private datasets collected by the authors for a specific clinical research purpose (figure 8.2).

8.2.1.2 Methods

Some general methods directly apply one or several ensemble deep convolutional neural networks for the classification of DR. Ensemble models are considered to improve the generalization and robustness and are widely used in some challenges

Table 8.1. DR grading descriptions.

Grade	Description
Normal	No visible lesions
NPDRI	Only microaneurysms
NPDRII	Microaneurysms, hemorrhages, and/or exudates
NPDRIII	More than 20 hemorrhages
PDR	Neovascularization

Table 8.2. Widely used public datasets for diabetic retinopathy recognition.

Dataset	Number of images	Number of grades
IDRiD[1]	516	5
Messidor[2]	1200	4
Messidor-2[3]	1748	5
DDR[4]	13 673	6 (with an ungradable category)
EyePACS[5]	35 126 train, 53 576 test	5

[1] https://ieee-dataport.org/open-access/indian-diabetic-retinopathy-image-dataset-idrid
[2] http://www.adcis.net/en/third-party/messidor/
[3] http://www.adcis.net/en/third-party/messidor2/
[4] https://github.com/nkicsl/DDR-dataset
[5] https://www.kaggle.com/c/diabetic-retinopathy-detection/data

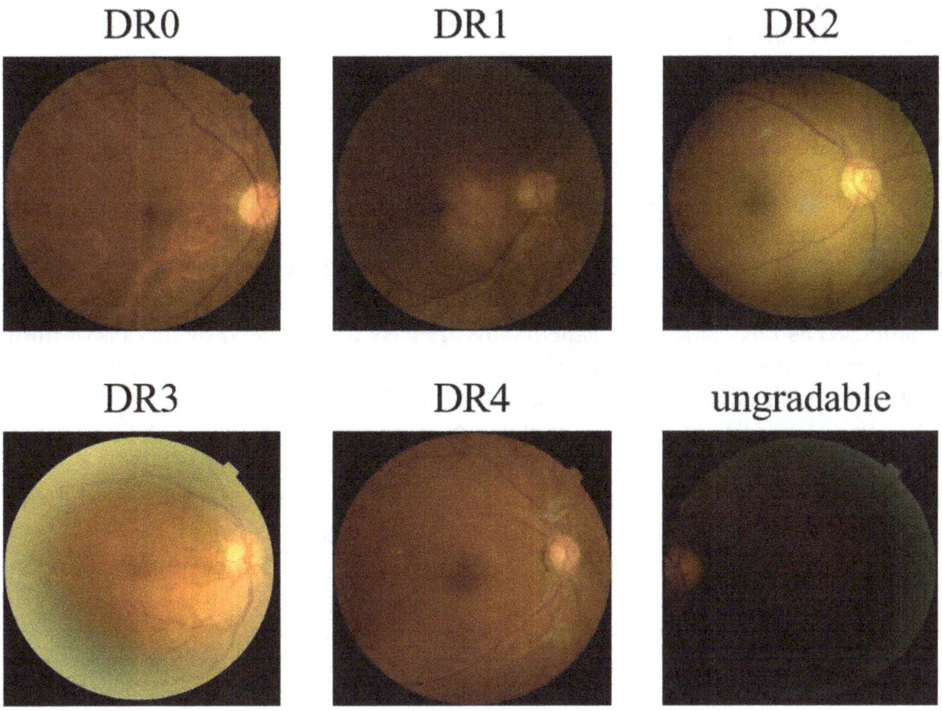

Figure 8.2. Examples from the DDR dataset. We use DR0–4 to refer to the different stages described in table 8.1.

and clinical papers. Gulshan *et al* [6] designed a framework based on Inception-V3 architecture [7] for the purpose of giving a diagnosis of with or without DR. It conducted the experiments on two datasets, EyePACS and Messidor, and the overall 90.00% area under the ROC curve (AUC) is achieved from the ensemble outputs of all trained networks. Krause *et al* [8] further improved the network selection and replace the Inception-V3 with Inception-V4, which achieved reliable results on the grading of 5 levels of DR.

Table 8.3. The grade distribution of the three public datasets.

Datset	DR0	DR1	DR2	DR3	DR4
Messidor	45.6%	12.7%	20.6%	21.1%	NA
EyePACS	73.3%	6.9%	15.2%	2.6%	2.0%
DDR	48.9%	5.1%	36.6%	1.9%	7.5%

There are some other works focused on the binary classification of normal/DR [9–11]. In particular, Wang *et al* [10] proposed a multi-scale network called Zoom-in-Net, which prevents the low-resolution input from ignoring the small lesions such as microaneurysms and blot hemorrhages by obtaining a crop of the retinal images. Pour *et al* [11] leveraged the CLAHE method to increase the contrast of input images and also improved the sensitivity of the model to those referable DR samples.

Compared with binary classification, the grading of different stages of DR can be more challenging. There are several reasons for this. First, most public and private datasets are imbalanced. As table 8.3 shows, DR0 and DR2 occur more frequently than others. The imbalanced distribution makes the model tend to pay less attention to those minority classes and does impact the generalization of the model. For this, He *et al* [12] tackled this as a fine-grained challenge and proposed a novel category attention block (CAB) to treat each grade of DR more equally. Second, the boundary between some categories is not clear and intra-observer variability is introduced, e.g. DR1 can be misdiagnosed as DR0 since microaneurysms are easy to be missed by models or human experts particularly under a low-resolution input. Ju *et al* [13] re-labeled the EyePACS dataset, which is estimated to have more than 20% mislabeled samples in the training set. For this, Araújo *et al* [14] proposed a system named DR|GRADUATE and give a measurement on how uncertain the predictions are. Also, the multiple instance learning (MIL)-based methods introduce more interpretability to the model. A few recent works [15–17] denote the importance of DR grading for the early treatment of vision loss.

To improve the interpretability of the model, some works leveraged the internal relationship between lesions and grades. Yang *et al* [18] pointed out the positions of the lesions by generating heatmaps from making a diagnosis on the stages of DR. Some datasets contain both annotations on grading and lesions and can be used for multi-task learning. Recently, Foo *et al* [19] applied the segmentation of lesions and DR grading at the same time, which can further improve the performance of the two branches.

8.2.1.3 Discussion

The DR classification was primarily explored for retinal disease recognition and has been studied widely in both the clinical and engineering fields. Most works present satisfactory results under a controlled environment. However, there still are some challenges and potential directions in future research:

1. The imbalance has a catastrophic impact on the generalization of the model to minority classes such as NPDRI and PDR.

2. The grading of DR sometimes indicates how serious the condition of a patient is. Regarding it as a regression problem may help reduce the risk of misdiagnosis between similar levels with unclear boundaries.
3. The grading of DR, which is determined by what and how many lesions are in present in the retina, should be a reasoning process. Since the grading scale is known, the final prediction can be made if we are able to quantify the lesions.
4. The model is expected to give not only a final prediction but also its basis. Current works can only tell where the lesions are but lack more detailed explanations. Releasing more datasets with lesion-level annotations can help tackle this issue.

8.2.2 Age-related macular degeneration

Age-related macular degeneration (AMD) is a leading cause of seriously impaired vision and usually occurs when people are in their 50s and 60s. As reported by [20], AMD accounts for 8.7% of legal blindness worldwide. For AMD grading, there is no unified standard, and it can be divided roughly into different stages according to the physical signs. For example, drusens can be classified according to size and appearance [21, 22], as figures 8.3(a) and (b) show. Figure 8.3(c) shows the development of wet AMD with a subfoveal hemorrhage surrounding it, resulting from a large serous detachment of the retinal pigment epithelium (RPE). Figure 8.3(d) shows another typical wet AMD which occurs due to geographic atrophy (GA).

8.2.2.1 Datasets

Compared with DR classification, there are not so many public datasets for AMD. We have summarized three widely used AMD datasets with at least 1000 samples in table 8.4. Also, in the iChallenge-AMD dataset the annotations of disc boundaries and fovea locations, as well as the boundaries of the different types of lesions are provided for further assessment.

8.2.2.2 Methods

Govindaiah *et al* [23] proposed a VGG-16-based [24] framework to achieve the classification of 2/4 stages of AMD evaluated on AREDS. The authors also used

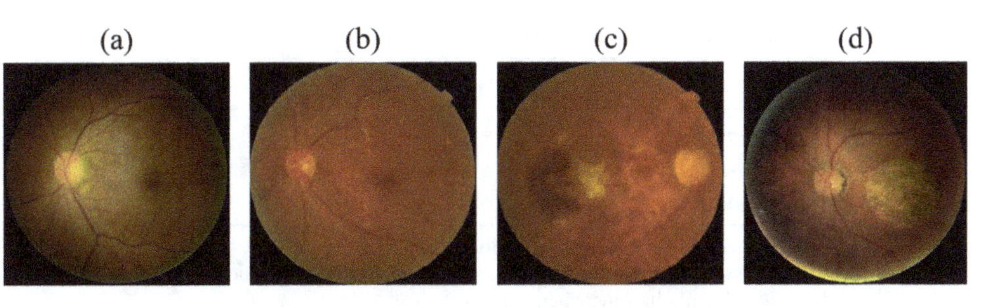

Figure 8.3. Examples of different stages of AMD.

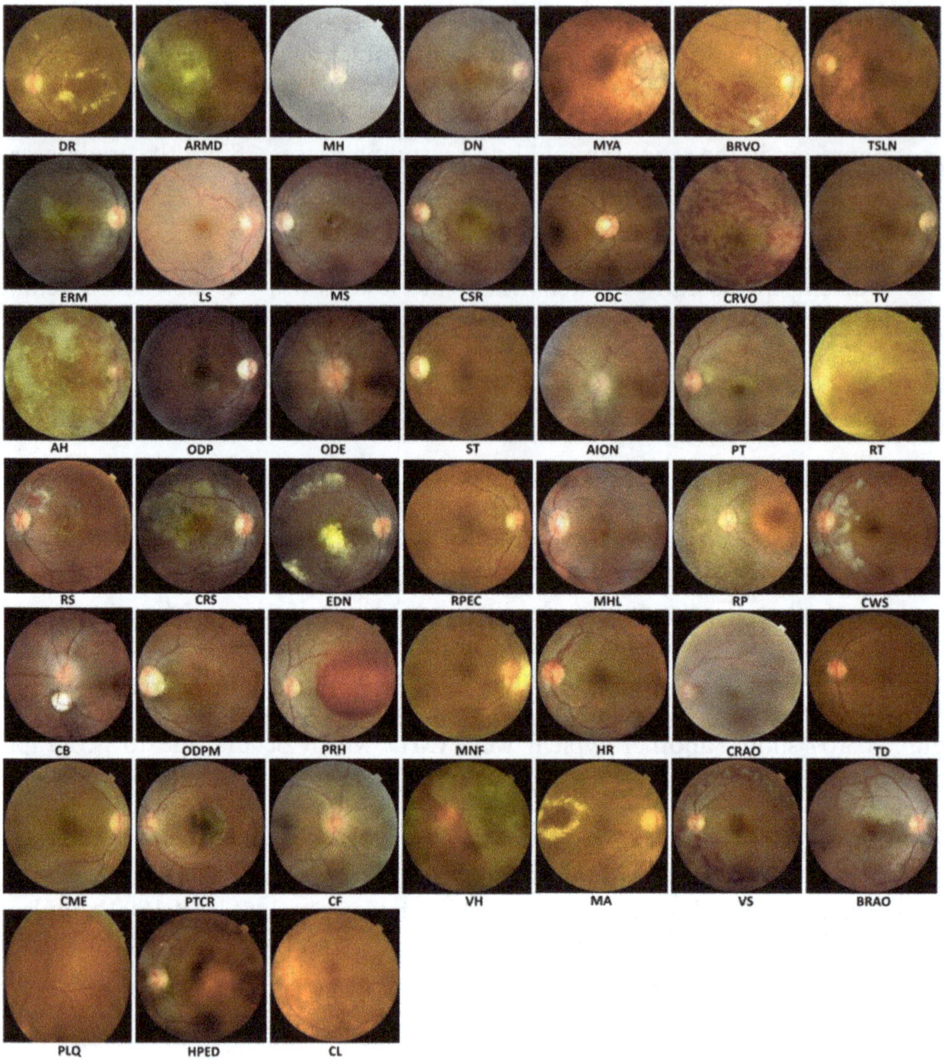

Figure 8.4. Examples of images from each targeted condition from RFMiD [3].

Table 8.4. Widely used public datasets for AMD recognition.

Dataset	Number of images
iChallenge-AMD[1]	1200
AREDS[2]	20000+
KORA[3]	2000+

[1] http://ai.baidu.com/broad/introduction?dataset=amd
[2] https://www.ncbi.nlm.nih.gov/projects/gap/cgi-bin/study.cgi?study
[3] https://epi.helmholtz-muenchen.de/

pre-processing to eliminate the noisy samples resulting from lens glare and non-uniform illumination. Grassmann *et al* [25] proposed an ensemble network consisting of six different convolutional neural network (CNN) architectures. Then, the features extracted from different networks were fused into a random forest model for 13 stages of AMD classification. As we declared in section 8.2.1.3, the accuracy of AMD prediction can be improved by the extra supervision information such as the annotations of the positions of corresponding lesions. Peng *et al* [26] proposed a novel framework that consists of three streams, which can be used for different stages of grading and lesions detection.

8.2.2.3 Discussion

It can be seen that compared with DR, the study of automated diagnosis of AMD has received less attention. In fact, the physical signs are much more complex and various, but there is no work that makes a comprehensive analysis of them using AI-based technology. Although some late AMD stages can be detected easily due to their obvious extensive degeneration, some early stages can be confused with other non-referable cases such as pigmentation or other diseases such as hard exudates. A small drusen can be missed in the early stage, which causes a delay in treatment. We hope more newly released datasets in the future can attract more researchers.

8.2.3 Other common retinal diseases

8.2.3.1 ROP

Retinopathy of prematurity (ROP), caused by the disorganized growth of retinal blood vessels [27] in childhood, is one of the leading causes of blindness for children. Zhang *et al* [28] proposed an approach with two DNN models to automatically diagnose AP-ROP (Aggressive posterior ROP). Specifically, the first network identifies whether the image has the presence of ROP, and the second network divides the ROP images into regular ROP and AP-ROP. Peng *et al* [29] proposed a network that used pre-trained ResNet18 as the backbone and added an attention block called the complementary residual attention block (CRAB) and equeeze-and-excitation (SE) block. They finally achieved a 99.84% AUC by evaluating 9794 fundus images. Hu *et al* [30] trained two sub-networks connected by a feature aggregate operator, which achieved high classification accuracy on the prediction of stages.

Some works [31–33] attempt to learn from case-level annotations via MIL. MIL shows advantages when there is a limited number of labeled images. Also, it can improve the detection accuracy of the subtle degeneration in the early stage. Rani *et al* [32] split the fundus image into a grid of patches, and instances are constructed from each grid element by extracting a set of features from it. All the feature sets or groups of instances belonging to the same image are grouped into a bag, and the bags along with their labels are fed into an MIL classifier for classification. Li *et al* [33] proposed an MIL-based framework that embedded a spatial attention (SA) module into the MIL mechanism. This work achieved case classification, instance selection, and abnormality localization for ROP in a unified framework.

There are still some issues that need to be addressed. First, the public datasets for ROP are few, and there is not enough attention on such a common retinal disease of children. Second, the degeneration of early ROP is tiny, even almost invisible to human experts, so some works have considered pre-processing techniques to enhance the low contrast but the detection accuracy is still very limited. Third, most works focus on the grading of stages in a small range of retina due to the limited imaging range of the commonly used RetCam. For better analysis of ROP in a global perspective, ultra-widefield fundus cameras should be introduced.

8.2.3.2 Cataracts

A cataract is a clouding of the normally clear lens of the eye. In real clinical practice, fundus imaging may not be the best way to diagnose cataracts, particularly for automated screening by an AI model. Because the operation is not standard, some images can be blurred and these can easily be misdiagnosed as cataracts. There are several works that explore the classification of cataracts. Zhou *et al* [34] used the improved Haar features and visible structure features, then a multilayer perceptron is followed as classifiers, which can grade four stages of cataracts. Xu *et al* [35] proposed to learn useful features directly from input data, and the de-convolution network method is employed to investigate how CNN characterizes the cataract layer-by-layer.

When it applied to a practical application, it is necessary for future works to distinguish between fundus images blurred by imaging errors and cataracts. Also, there are no unified grading standards and public datasets.

8.2.4 The challenge of multiple retinal disease recognition

As we discussed above, deep learning-based methods have proven to be effective for recognizing one or two kinds of common retinal diseases or pathologies under some controlled experimental environments. However, one patient may suffer from multiple kinds of retinal diseases and a single model may lose the ability of globally monitor the condition of the patient. What is more, deep learning is data-driven and the model tends to be less suited to those rare diseases without enough examples for training [36, 37]. Thus, the model will also ignore those rare conditions.

8.2.4.1 Datasets

The datasets for multiple diseases require more samples collected from different scenarios and human experts are needed for labeling. Hence, such datasets are extremely scarce. The ODIR[1] dataset consists of 10 000 paired fundus images from 5000 patients but there are only patient-level annotations on eight different retinal diseases, such as DR, glaucoma, cataract, etc. RFMiD [38] consists of 3200 fundus images with 46 conditions, including common diseases and some rare diseases, as figure 8.4 shows. Ju *et al* [13] released a dataset that re-labeled the Kaggle-EyePACS dataset as Kaggle DR+. The newly released dataset was extended into more than 50

[1] https://github.com/nkicsl/OIA-ODIR

conditions. Kaggle DR+ maintains the crowdsourcing annotations from multiple experts for a better study of the disagreement and observer variability from each expert when giving opinions. It is also accompanied by a golden standard database that has unbiased annotation after discussion and can be leveraged for validation.

8.2.4.2 Methods

In this section, we first discuss the works on simultaneous diagnosis of two or three kinds of retinal diseases. Most works focus on a multi-task for DR and diabetic macular edema (DME). Li *et al* [39] designed a novel network named CANet with two different types of attention models for the specific diseases. Tu *et al* [40] proposed a feature separation and union network (SUNet) for simultaneous DR and DME grading. This framework aimed to discover irrelevant features and leverage the features of different but related tasks to improve the performance of each sub-task. Similarly, Ju *et al* [41] proposed synergic adversarial label learning (SALL) for sharing information between two tasks such as DR and AMD. The SALL system learns to find similar information in the feature space by generating pseudolabels. Then, knowledge distillation is leveraged for learning better from the soft pseudo-labels. Ting *et al* [42] designed a network for simultaneous DR, glaucoma, and AMD diagnosis. They evaluated their methods on a dataset with about 500 000 samples.

Recently, some researchers began to explore automated screening for more kinds of retinal diseases with some released datasets such as ODIR, which contains eight categories. Li *et al* [43] proposed a dense correlation network (DCNet) to exploit the dense spatial correlations between the paired color fundus photographs. The DCNet contains a spatial correlation module that can capture the dense correlations between the features extracted from the paired color fundus photographs in a pixel-wise manner, and fuse the relevant feature representations. Wang *et al* [44] first designed a three-stream framework that can detect 36 kinds of retinal diseases. The framework first leveraged the YOLO-V3 [45] to detect regions such as the optic disc and macular area. Then, three branches are followed for the detection of specific diseases in different regions in a multi-task manner. Quellec *et al* [36] used a few-shot learning framework with an unsupervised probabilistic model for rare condition detection. They performed experiments on the OPHDIAT dataset with 41 conditions and 37 conditions can be detected with an AUC greater than 0.8. Ju *et al* [37] considered the multiple retinal diseases diagnosis to be a multi-label long-tailed classification challenge. In this work, the original long-tailed distribution is divided into several subsets according to the similarity or occurred regions. Then, subsets are used to train several teacher models whose information will be distilled into a unified student model in the following stage. They selected ten kinds of lesions and 48 kinds of diseases as target tasks and achieved competitive results compared with some SOTA methods.

8.2.4.3 Discussion

As two datasets released by Pachade *et al* [38] and Ju *et al* [37] show, the distribution of retinal diseases always exhibits a long-tailed attribute since some conditions are common and some are rare (see figure 8.5). One issue is that the samples for those rare cases are difficult to collect. The deep learning model shows less robustness to those minority classes. The re-sampling strategies are potential solutions for long-tailed classification. However, there exists a label co-occurrence in retinal diseases, i.e. one sample contains more than one condition. The re-sampling for those minority classes may lead to over-sampling on those majority classes, which brings a new relative imbalance issue to the model training.

Another challenge is how to leverage the cognitive laws or prior knowledge in ophthalmology. For instance, there are sub-types in most diseases such as DR and glaucoma. So retina diseases can be organized in a hierarchical manner. Embedding that prior knowledge in an AI model may help in better understanding the semantic information or learning features among so many categories in such a complex medical system. Also, a coarse-to-fine prediction can bring more useful information for the global monitoring of the condition of patients (figure 8.6).

A similar challenge to hierarchical classification is multi-label learning. In the classification of natural images, some methods [46] aim to explicitly exploit the label dependencies in an image, e.g. the sky is always accompanied by clouds. However, there is no related work in multiple retinal disease recognition.

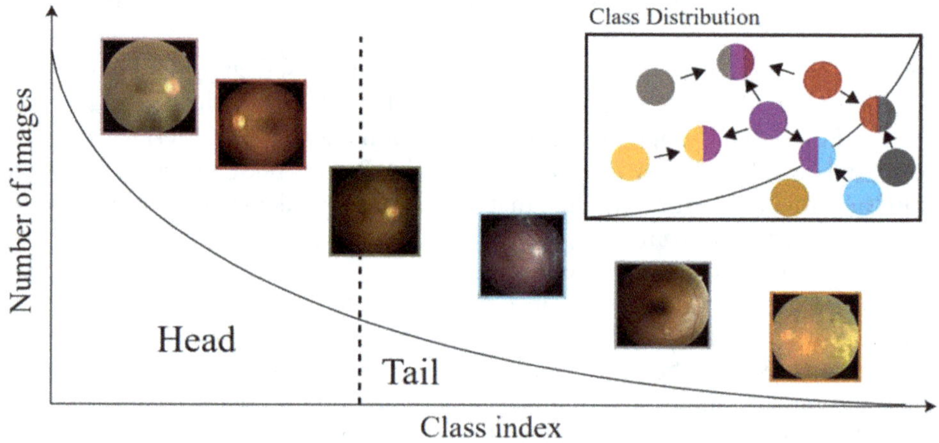

Figure 8.5. The retinal disease distribution exhibits a prominent long-tailed attribute [37].

Main Category	Sub Category
normal	normal
myopia	myopia, path_myopia_miman
DR	NPDRI, NPDRII, NPDRIII, PDR
AMD	macula_small_drusen, macula_big_drusen, macula_GA, macula_CNV
macula	macula_other, macula_edm, macula_csc, macula_em, macula_hole
hypertension	less_elatic_artery, hypertensionI, hypertensionII, hypertensionIII, BRVO, BRAO
other	other, RP, amotio, VKH, VH, VD, Coats, retina_aneurysm, choroidal_osteoma,
optic disc	disc_other, disc_Myel_nerve, disc_atrophy, disc_edm, disc_glaucoma,
	disc_glory, disc_melanocytoma
	choroidal_hemangioma, si_eye, laserpoint, choroid_coloboma, peripheral_degeneration

Figure 8.6. The hierarchies of retinal diseases proposed in [13].

The diagnosis of multiple diseases is very significant in clinical practice and deserves more attention. We hope more datasets can be released in the future and more experiments can be carried on public or private datasets.

8.3 Segmentation

8.3.1 Lesion segmentation

8.3.1.1 Introduction

Compared to the task of classification and detection, retinal lesion segmentation is a more precise and detailed representation of the type and area of fundus lesions at the pixel level of granularity. It has crucial applications in tasks such as quantification of the lesion region and quantitative tracking of disease progression, which are time-consuming and challenging to measure accurately with a manual approach. A wide variety of lesions can be observed on fundus images, such as drusen, hard exudate (EX), soft exudate (SE), microaneurysm (MA), hemorrhage (HE), epiretinal membrane, macular hole, and retinitis pigmentosa, etc. It is the basis for the diagnosis of retinal diseases using fundus images. These lesions vary in size and shape, and some of the categories are very similar and easily confused, making the task difficult and placing higher demands on the algorithm design. We are pleased to see that several datasets for fundus image lesion segmentation have been proposed in recent years, which provides good support for the advancement of technology in this area.

8.3.1.2 Datasets

The images in the retinal lesion segmentation datasets are generally limited compared to the segmentation tasks in natural scenes due to the difficulty of data acquisition and the high cost of labeling. Meanwhile, although there are many types of fundus lesions, the ones that have been much studied in segmentation tasks are mainly those related to the diabetic retina, i.e. EX, SE, MA, HE, etc.

Table 8.5 lists the details of some of the most commonly used retinal lesion segmentation datasets. It can be seen that except for the DDR dataset, the sample size of all the other datasets is around one hundred images. The lack of accessible large-scale retinal lesion segmentation datasets may also be an obstacle in this research area. Hence, building an efficient network to cope with small sample

Table 8.5. Details of commonly used retinal lesion segmentation datasets.

Database	Year	No. of images	Pixel level annotation?	MA	EX	SE	HE
E-ophtha_EX [47]	2013	47	Yes		✓		
E-ophtha_MA [47]	2013	148	Yes	✓			
IDRiD [48]	2018	81	Yes	✓	✓	✓	✓
DDR [49]	2019	757	Yes	✓	✓	✓	✓
DIARETDB1 [50]	2007	89	No	✓	✓	✓	✓

Figure 8.7. Example of the DDR dataset. The original image and ground truth for MA, EX, SE, and HE are shown from left to right.

datasets is also a requirement for the lesion segmentation task. Figure 8.7 shows a sample of lesion segmentation in the DDR dataset.

8.3.3.3 Methods

The study of fundus lesion segmentation has a long history. In the early studies, image processing-based methods were usually used. Baudoin *et al* [53] and Walter *et al* [54] used morphology-based methods to detect microaneurysms. These early methods were mainly based on some primary features, such as shape, color, or texture, while the variable shape of lesions in the fundus, together with the influence of image illumination and hue variations, made it difficult to accurately segment the lesions and did not generalize well. Thanks to the rapid development of artificial neural network technology in the last decade, fundus lesion segmentation techniques have also made great progress. Specifically, the current techniques for fundus lesion segmentation are adaptations of generic segmentation methods on fundus images.

To cope with the difficulty of lesion segmentation, several targeted methods or techniques have been designed or introduced for the task of lesion segmentation of fundus images. Guo *et al* [51] proposed a segmentation network (L-Seg) for small targets for the characteristics of retinal lesions and used a multi-scale feature fusion strategy to alleviate the problem that the features of small targets are difficult to express in a high-level network. Also, multi-channel bin loss is proposed considering the problem of unbalanced lesion categories. Similarly, Xu *et al* [58] introduced multi-scale feature fusion and contextual channel attention to the U-Net network to enhance its information fusion capability, and used improved focal loss to alleviate the data imbalance. Xue *et al* [52] combined the membrane computation technique with mask R-CNN [55] to bring better convergence, robustness, and parallelism to fundus lesion segmentation.

There are also approaches that try to combine lesion segmentation with other related tasks that make fuller use of image information or information from different modalities. Wang *et al* [56] constructed a multi-tasking network that uses super-resolution and lesion segmentation to assist in diabetic retinal grading, and uses feedback from segmentation and grading to guide the pathological regions of super-resolution task focus. Zhou *et al* [57] constructed a cross-modal network that attempts to extract information from the diagnostic text to assist in fundus lesion segmentation.

8.3.3.4 Evaluation and results

In tables 8.6–8.8 we have collected the results of some SOTA methods on commonly used datasets, evaluated by the most widely used segmentation evaluation metrics, including the area under the ROC curve (AUC), average precision (AP), accuracy (ACC), specificity (SP), sensitivity (SE), and the DICE score.

Table 8.6. Existing SOTA methods on the IDRiD dataset.

Reference	Year/Journal	Task	AUC	ACC	SP	SE	DICE
Guo [51]	2019/*Neurocomputing*	HE	0.6734	—	—	—	—
Xu [58]	2021/*Biomed. Res. Int.*	HE	—	—	—	0.7342	0.8450
Guo [51]	2019/*Neurocomputing*	MA	0.4627	—	—	—	—
Xue [52]	2019/*Knowl.-Based Syst.*	MA	—	0.997	0.998	0.764	—
Xu [58]	2021/*Biomed. Res. Int.*	MA	—	—	—	0.5933	0.7188
Guo [51]	2019/*Neurocomputing*	EX	0.7945	—	—	—	—
Xue [52]	2019/*Knowl.-Based Syst.*	EX	—	0.992	0.996	0.779	—
Xu [58]	2021/*Biomed. Res. Int.*	EX	—	—	—	0.8755	0.9138
Guo [51]	2019/*Neurocomputing*	SE	0.7113	—	—	—	—
Xu [58]	2021/*Biomed. Res. Int.*	SE	—	—	—	0.7933	0.8812

Table 8.7. Existing SOTA methods on the DDR dataset.

Reference	Year/Journal	Task	AUC	AP
Guo [51]	2019/*Neurocomputing*	EX	0.5546	—
Wang [56]	2021/*Proc. AAAI*	EX	0.990	0.718
Guo [51]	2019/*Neurocomputing*	SE	0.2648	—
Wang [56]	2021/*Proc. AAAI*	SE	0.992	0.647
Guo [51]	2019/*Neurocomputing*	HE	0.3586	—
Wang [56]	2021/*Proc. AAAI*	HE	0.971	0.554
Guo [51]	2019/*Neurocomputing*	MA	0.1052	—
Wang [56]	2021/*Proc. AAAI*	MA	0.987	0.501

Table 8.8. Existing SOTA methods on the E-ophtha databset.

Reference	Year/Journal	Task	AUC	ACC	SP	SE
Guo [51]	2019/*Neurocomputing*	MA	0.1687	—	—	—
Xue [52]	2019/*Knowl.-Based Syst.*	MA	—	0.997	0.998	0.672
Guo [51]	2019/*Neurocomputing*	EX	0.4171	—	—	—
Xue [52]	2019/*Knowl.-Based Syst.*	EX	—	0.984	0.988	0.846

8.3.3.5 Summary

These existing methods reflect the exploration by researchers in different solution directions. Due to the characteristics of fundus lesions themselves and the limitations of insufficient current datasets, semi-supervised or unsupervised learning, multi-task learning, cross-modal collaborative learning, and multi-instance learning, which are current research hotspots, have potential applications in fundus lesion segmentation tasks. We are looking forward to fundus lesion segmentation being elevated to a higher level with the development of these techniques.

8.3.2 Vessel segmentation

8.3.2.1 Introduction

Retinal vessel segmentation plays a critical and fundamental role in constructing a computer-aided diagnosis (CAD) system. The quantitative analysis of retinal vessels can provide the bases, including the length, tortuosity, and diameter, of retinal vessels for fundus image processing, e.g. retinal artery/vein classification [59], fundus registration [60], and biomedical identification [61]. In this way, such a CAD system can be further used to automatically screen or diagnose some related fundus diseases such as diabetic retinopathy (DR), glaucoma, and cataracts, which can relieve the heavy burden on clinicians effectively [62].

8.3.2.2 Datasets

Here we use four popular public databases, including DRIVE [63], STARE [64], CHASE_DB1 [65], and HRF [66], to evaluate the existing segmentation methods. The details of these four databases are shown in table 8.9. The example retinal images on the DRIVE database are shown in figure 8.8. Note that field of view (FOV) masks are given on the DRIVE and HRF databases, which can be used to indicate the region of interest inside the retinal images. Hence, we can manually generate the FOV masks of the STARE and CHASE_DB1 databases as in [67].

Table 8.9. Details of four retinal vessel segmentation databases.

Database	No. of images	Resolution	Train/Test	FOV
DRIVE	40	565×584	20/20	Yes
STARE	20	700×605	10/10 or 4-fold	No
CHASE_DB1	28	999×960	20/8 or 4-fold	No
HRF	45	3504×2336	15/30 or 30/15	Yes

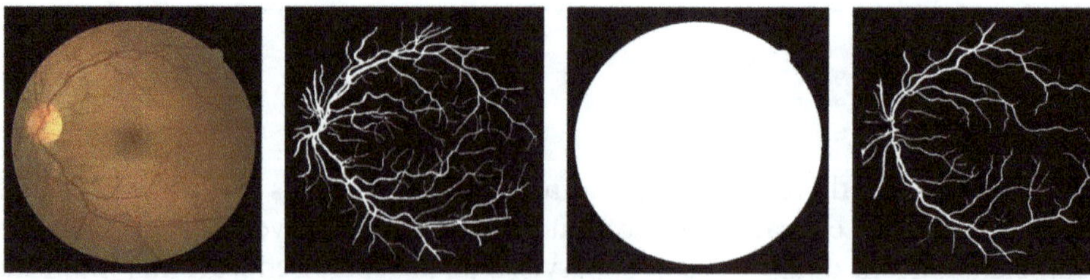

Figure 8.8. Examples from the DRIVE database. The original image, ground truth, mask, and U-Net prediction are shown from left to right.

8.3.2.3 Methods

Retinal vessel segmentation is a hot topic in the field of medical image processing. Recent advances of retinal vessel segmentation are mainly based on deep learning technologies. We can roughly divide the existing state-of-the-art (SOTA) methods into three categories as in the following.

8.3.2.3.1 'Divide and conquer' strategy

The first category is based on the 'divide and conquer' strategy. There are some differences between thin retinal vessels and thick retinal vessels. Dividing the multi-scale vessels can decrease the inter-class differences. For example, Yan *et al* [68] proposed a three-stage deep learning framework to complete thick blood vessel segmentation, thin blood vessel segmentation, and blood vessel fusion to minimize the imbalance of positive and negative pixels. Wang *et al* [69] designed a network composed of three decoders called HAnet. The first decoder analyzes whether the location region is 'hard' or 'easy', while the other two decoders segment 'hard' and 'easy' regions independently. In addition, an attention mechanism is introduced to enhance the features of 'hard' regions. Zhang *et al* [70] proposed a structure–texture demixing network (STD-Net) to separate the structure and texture of the retinal image and process them separately to obtain better segmentation performance.

8.3.2.3.2 Multi-scale methods

Since the thin and thick retinal vessels vary a lot, many deep learning-based methods are proposed to extract more effective multi-scale features to accurately detect retinal vessels. For instance, Wu *et al* [71] designed an efficient Inception–Residual convolutional block to combine the advantages of the Inception model and Residual module for improved feature representation and embedded the residual block into the U-like encoder–decoder for vessel segmentation. Wang *et al* [72] proposed a dual encoding U-Net (DEU-Net), which contains two encoders. One employs a large-scale convolution kernel to retain more spatial information. The other employs multi-scale convolution blocks to more more contextual semantic information. At the end of the two encoders, a feature fusion module is used to fuse spatial information and context information. Xu *et al* [73] enhanced the U-shaped network by introducing a recursive semantics-guided network, which integrates rich semantic information into a shallow layer to obtain detailed information. Wang *et al* [74] proposed a novel feature pyramid cascade network (RVSeg-Net) for retinal vessel segmentation to alleviate the problems of large vessel size change, a large number of capillaries, and discontinuity. RVSeg-Net introduces a feature pyramid cascade (FPC) module to capture multi-scale information and aggregate local and global context information. In addition, octave convolution is employed instead of vanilla convolution to greatly reduce the model parameters. Wu *et al* [75] proposed a scale and context-sensitive retinal vessel segmentation network (SCS-Net). The SCS-Net contains three novel modules. The scale-aware feature aggregation (SFA) module is used to adjust the receptive field dynamically. The adaptive feature fusion (AFF) module is introduced to guide the fusion of adjacent features. Finally, the multi-level semantic supervision (MSS) module is employed to learn more distinctive semantic representation. Li *et al* [76] proposed a full attention network (FANet) based on an attention mechanism to adapt to learn rich features and integrate multi-scale information. The FANet includes a lightweight bidirectional attention module to introduce global dependencies to reduce intra class inconsistencies. At the same time, selective kernel (SK) units are used to replace vanilla convolution to obtain multi-scale features. Retinal vessel segmentation and centerline extraction were regarded as independent tasks in previous works, and the relationship between them is usually ignored. Therefore, Xu *et al* [77] proposed a deep semantics and multi-scaled cross-task aggregation network, which uses the association between two tasks to improve their performance.

8.3.2.3.3 Combing with traditional methods

Furthermore, the traditional image processing methods can be combined with deep learning-based methods to further improve the performance. For example, Mou *et al* [78] employed the dense expansion network to extract features with different scales and introduced a probability regularized walk algorithm to address the broken blood vessels, which improves the connectivity of blood vessels. Zhang *et al* [79] proposed a boundary enhancement and feature denoising (BEFD) module to improve the ability to extract boundary information in semantic segmentation. In this module, a Sobel edge detector is introduced to obtain additional edge prior information, and the denoising block is used to reduce the noise in low-level features.

8.3.2.4 Evaluation and results

In this area, the AUC, ACC, SP, SE, and the F1 score are usually used to evaluate the segmentation methods. At the same time, existing methods vary in their specific pre-processing methods and training/testing setting. In tables 8.10–8.13 we provide the results of some representative SOTA methods for four public databases, respectively. Note that most of methods are evaluated inside the FOV masks in the original size of the retinal images. Meanwhile, some models are evaluated in the whole image. Furthermore, the resizing operation is used to change the size of the retinal images. We labeled these three classes in each table.

8.3.2.5 Summary

The existing models have achieved great improvements in retinal vessel segmentation in terms of the above metrics. However, the connectivity or topology of retinal vessels cannot be preserved by such deep learning-based models. Meanwhile, existing models cannot segment capillaries well. Hence, new evaluation metrics and novel models may focus on these challenges in the future.

8.3.3 Optic disc/cup segmentation

8.3.3.1 Introduction

Glaucoma is one of the leading causes of irreversible blindness [80]. Early screening for glaucoma is very important. The cup-to-disc ratio (CDR) in fundus images is an important basis for the diagnosis and screening of glaucoma [81]. The larger the

Table 8.10. Existing SOTA methods on the DRIVE database. (MICCAI = *International Conference on Medical Image Computing and Computer-Assisted Intervention.*)

Evaluation	Reference	Year/Journal	AUC	ACC	SP	SE	F1	Train/Test
Inside the FOV masks	Yan [68]	2019/*IEEE J. Biomed. Health Inform.*	0.9750	0.9538	0.9820	0.7631	—	20/20
	Wang [69]	2020/*IEEE J. Biomed. Health Inform.*	0.9823	0.9581	0.9813	0.7991	0.8293	
	Xu [77]	2020/*IEEE J. Biomed. Health Inform.*	0.9821	0.9571	0.9750	0.8339	0.8319	
	Mou [78]	2020/*IEEE Trans. Med. Imaging*	0.9796	0.9594	0.9788	0.8126	—	
	Wu [71]	2019/MICCAI	0.9821	0.9578	0.9802	0.8038	—	
	Wang [72]	2019/MICCAI	0.9772	0.9567	0.9816	0.7940	0.8270	
	Xu [73]	2020/MICCAI	0.9810	—	0.9470	0.9120	—	
Resize image	Wang [74]	2020/MICCAI	0.9817	0.9681	0.9845	0.8107		
	Zhang [79]	2020/MICCAI	0.9867	0.9701	0.9845	0.8215	0.8267	
	Zhang [70]	2020/MICCAI	0.9863	0.9695	0.9846	0.8151	—	
	Wu [75]	2021/*Med. Image Anal.*	0.9837	0.9697	0.9838	0.8289	—	
Whole image	Li [76]	2020/*IEEE J. Biomed. Health Inform.*	0.9895	0.9769	0.9883	0.8145	—	

Table 8.11. Existing SOTA methods on the STARE database. (MICCAI = *International Conference on Medical Image Computing and Computer-Assisted Intervention.*)

Evaluation	Reference	Year/Journal	AUC	ACC	SP	SE	F1	Train/Test
Inside the FOV masks	Yan [68]	2019/*IEEE J. Biomed. Health Inform.*	0.9833	0.9638	0.9857	0.7735	—	Leave-one-out
	Wang [69]	2020/*IEEE J. Biomed. Health Inform.*	0.9881	0.9673	0.9844	0.8186	0.8379	Leave-one-out
	Xu [77]	2020/*IEEE J. Biomed. Health Inform.*	0.9881	0.9664	0.9802	0.8463	0.8384	5-fold
	Mou [78]	2020/*IEEE Trans. Med. Imaging*	0.9858	0.9685	0.9769	0.8391	—	4-fold
	Xu [73]	2020/MICCAI	0.9860	—	0.9540	0.9280	—	10/10
	Wu [75]	2021/*Med. Image Anal.*	0.9877	0.9736	0.9839	0.8207	—	10-fold
Whole image	Li [76]	2020/*IEEE J. Biomed. Health Inform.*	0.9924	0.9797	0.9889	0.8505	—	10/10

Table 8.12. Existing SOTA methods on the CHASE_DB1 database. (MICCAI = *International Conference on Medical Image Computing and Computer-Assisted Intervention.*)

Evaluation	Reference	Year/Journal	AUC	ACC	SP	SE	F1	Train/Test
Inside the FOV masks	Yan [68]	2019/*IEEE J. Biomed. Health Inform.*	0.9776	0.9607	0.9806	0.7641	—	20/8
	Wang [69]	2020/*IEEE J. Biomed. Health Inform.*	0.9871	0.9670	0.9813	0.8239	0.8191	20/8
	Xu [77]	2020/*IEEE J. Biomed. Health Inform.*	0.9873	0.9651	0.9780	0.8508	0.8270	4-fold
	Mou [78]	2020/*IEEE Trans. Med. Imaging*	0.9812	0.9637	0.9773	0.8268	—	4-fold
	Wu [71]	2019/MICCAI	0.9860	0.9661	0.9814	0.8132	—	20/8
	Wang [72]	2019/MICCAI	0.9812	0.9661	0.9821	0.8074	0.8037	20/8
	Xu [73]	2020/MICCAI	0.9850	—	0.9540	0.9230	—	18/10
Resize image	Wang [74]	2020/MICCAI	0.9833	0.9726	0.9836	0.8069		20/8
	Wu [75]	2021/*Med. Image Anal.*	0.9867	0.9744	0.9839	0.8365	—	20/8
Whole image	Li [76]	2020/*IEEE J. Biomed. Health Inform.*	0.9912	0.9803	0.9862	0.8334	—	14/14

Table 8.13. Existing SOTA methods on the HRF database.

Evaluation	Reference	Year/Journal	AUC	ACC	SP	SE	F1	Train/Test
Inside the FOV masks	Wang [69]	2020/*IEEE J. Biomed. Health Inform.*	0.9837	0.9654	0.9843	0.7803	0.8074	15/30
Resize image	Wu [75]	2021/*Med. Image Anal.*	0.9842	0.9687	0.9823	0.8114	—	30/15

CDR value, the higher the risk of glaucoma. Accurate optic disc and cup segmentation is a prerequisite for measuring the CDR. Manual glaucoma screening is not only subjective but also time-consuming and labor-intensive and therefore not suitable for large-scale screening scenarios. It is therefore essential to establish a computer-aided diagnostic (CAD) system to automatically and accurately screen and diagnose glaucoma [82].

8.3.3.2 Datasets

Here we list several widely used public datasets for optic disc and cup segmentation, including: Drishiti-GS [109], REFUGE [103], RIM-ONE-r3 [108], and ORIGA [107]. Detailed statistical information for all datasets is shown in table 8.14. An example image from Drishiti-GS is presented in figure 8.9. Note that in order to better segment the optic cup and disc and avoid the influence of other regions of the fundus, the region of interest can be extracted from the original fundus image by a lightweight neural network for subsequent segmentation [103].

8.3.3.3 Methods

Optic disc and cup segmentation is a key step in fundus image analysis. Currently available optic disc and cup segmentation methods can be divided into two main categories: traditional methods and deep learning methods.

The traditional methods mainly extract features manually and then perform segmentation. For example, Lowell *et al* [83] used an active contour model to detect the optic disc boundary and Aquino *et al* [84] used a circular transform-based method to segment the optic disc. Cheng *et al* [85] used super-pixel point

Table 8.14. Details of four optic disc and cup segmentation databases.

Database	No. of images	Resolution	Camera
Drishiti-GS	101	2896×1944	Unknown
REFUGE	1200	2124×2056 1634×1634	Zeiss Viscucam 50/Canon CR-2
RIM-ONE-r3	159	1072×1424	Canon EOS 5D
ORIGA	650	3072×2048	Unknown

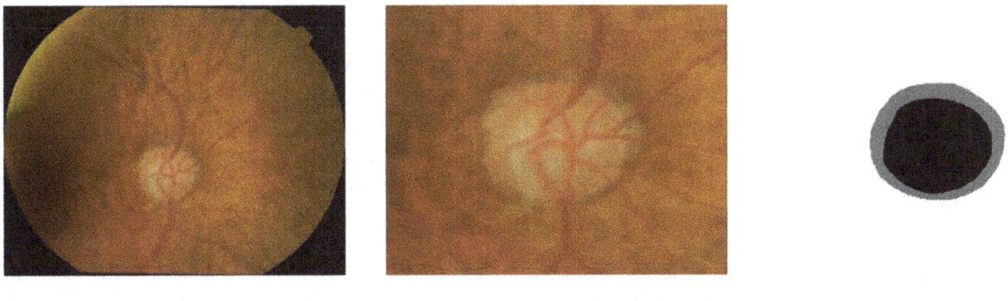

Figure 8.9. Example from the Drishiti-GS database. The original image, region of interest (ROI), and mask are shown from left to right.

classification to segment the optic disc and cup, transforming the edge detection problem into a pixel-by-pixel classification problem. Wong *et al* [86] used a level-set approach to segment the optic disc and completed the segmentation of the optic cup boundary based on the experience that blood vessels bend when crossing the optic cup boundary. The segmentation performance of traditional methods relies heavily on how well the image features are extracted manually, and the generalization performance of traditional methods is often poor.

Deep learning methods based on convolutional neural networks can automatically extract image features, and many deep learning-based networks for segmenting the optic disc and cup have been proposed. Mohan *et al* [87] proposed a cascaded neural network for segmenting the optic disc and cup. Gu *et al* [88] proposed the context encoder network (CE-Net), which constructed a context encoding module based on atrous convolution and spatial pyramid pooling to retain spatial and high-level information. Maninis *et al* [82] proposed deep retinal image understanding (DRIU) using a pre-trained VGG network as a feature extraction layer to perform both optic disc/cup segmentation and vascular segmentation. Al-Bander *et al* [89] used a fully convolutional DenseNet to segment the optic disc and cup simultaneously. Shankaranarayana *et al* [90] and Jiang *et al* [91] both based their approaches on generative adversarial networks to segment the optic disc and cup. Yu *et al* [92] proposed an modified U-Net and used a pre-trained ResNet34 to extract image features. Zhang *et al* [93] introduced an edge-attention module focusing on learning edge feature information from shallow feature maps and using the learned edge information to guide the deeper layers for segmentation. Xu *et al* [94] proposed a U-shaped segmentation network in combination with spatial pyramid to learn multi-scale features. Wang *et al* [95] treated the segmentation problem as a regression problem and directly fitted the positions of the optic disc and cup with ellipses. Jiang *et al* [96] add the prior knowledge that the optic cup is in the internal region of the optic disc to the network to constrain the segmentation results of the optic cup. Murugesan *et al* [97] design a multi-task network that outputs optic disc and cup segmentation maps, contour maps, and distance maps, respectively, using contour and distance maps to constrain the segmentation results. All of the above methods perform segmentation in a Cartesian coordinate system, and there are also some studies [98, 99] that perform optic disc and cup segmentation in a polar coordinate system.

8.3.3.4 Evaluation and results

The most widely used metrics for evaluating the performance of optic disc and cup segmentation are the dice coefficients (DI), intersection over union (IoU), means balanced accuracy (A in the tables), overlapping error (E in the tables), and absolute CDR error. In tables 8.15–8.18 we present the segmentation results of the currently available methods for each dataset.

8.3.3.5 Summary

Existing methods have achieved promising performance in the task of segmenting the optic disc and cup. However, currently existing methods typically suffer from

Table 8.15. Existing SOTA methods on the Drishiti-GS database. (MICCAI = *International Conference on Medical Image Computing and Computer-Assisted Intervention.*)

Reference	Year/Journal	OD		OC		Absolute CDR error
		Dice	IoU	Dice	IoU	
Edupuganti [100]	2018/*IEEE Int. Conf. Image Process.*	—	0.6958	—	0.8122	—
Mohan [101]	2018/*IEEE Int. Conf. Image Process.*	0.9640	—	—	—	—
Mohan [87]	2019/*IEEE Int. Conf. Image Process.*	0.9713	—	—	—	—
Liu [99]	2019/*Neurocomputing*	0.9800	—	0.8900	—	—
Shah [102] (PSBN)	2019/*Int. Symp. Biomed. Imaging*	0.9500	0.9100	0.8800	0.8000	—
Shah [102] (WroIM)	2019/*Int. Symp. Biomed. Imaging*	0.9600	0.9300	0.8900	0.8000	—
Wang [103]	2019/*IEEE Trans. Med. Imaging*	0.9740	—	0.9010	—	0.048
Wang [104]	2019/MICCAI	0.9610	—	0.8620	—	—

Table 8.16. Existing SOTA methods on the ORIGA database. (MICCAI = *International Conference on Medical Image Computing and Computer-Assisted Intervention.*)

Reference	Year/Journal	OD		OC		Absolute CDR error
		A	E	A	E	
Liu [98]	2019/*Neurocomputing*	—	0.059	—	0.208	—
Fu [98]	2018/*IEEE Trans. Med. Imaging*	0.9830	0.071	0.9300	0.230	0.071
Ying [105]	2019/MICCAI	0.9860	0.066	0.9420	0.208	0.065
Jiang [96]	2019/*IEEE Trans. Biomed. Eng.*	—	0.063	—	0.209	0.068

Table 8.17. Existing SOTA methods on the RIM-ONE-r3 database. (MICCAI = *International Conference on Medical Image Computing and Computer-Assisted Intervention.*)

Reference	Year/Journal	OD		OC		Absolute CDR error
		Dice	IoU	Dice	IoU	
Shah [102] (PSBN)	2019/*Int. Symp. Biomed. Imaging*	0.9100	0.8400	0.7500	0.6000	—
Shah [102] (WroIM)	2019/*Int. Symp. Biomed. Imaging*	0.9400	0.9000	0.8200	0.7100	—
Wang [103]	2019/*IEEE Trans. Med. Imaging*	0.9680	—	0.8560	—	0.049
Wang [104]	2019/MICCAI	0.8980	—	0.8100	—	—

Table 8.18. Existing SOTA methods on the REFUGE database. (MICCAI = *International Conference on Medical Image Computing and Computer-Assisted Intervention.*)

Reference	Year/Journal	OD			OC			Absolute CDR error
		A	E	Dice	A	E	Dice	
Yin [105]	2019/MICCAI	0.9790	0.088	—	0.9800	0.223	—	0.0480
Wang [95]	2019/*Int. Symp. Biomed. Imaging*	—	—	0.9530	—	—	0.8720	0.0470
Wang [103]	2019/*IEEE Trans. Med. Imaging*	—	—	0.9602	—	—	0.8826	0.0450
Liu [106]	2019/MICCAI	—	—	0.9416	—	—	0.8627	0.0481

degraded generalization performance when domain shifts are encountered. Therefore, further research into domain adaptation methods is needed to address this issue in the future.

References

[1] Abràmoff M D, Garvin M K and Sonka M 2010 Retinal imaging and image analysis *IEEE Rev. Biomed. Eng.* **3** 169–208

[2] Zhang Z, Srivastava R, Liu H, Chen X, Duan L, Wong D W, Kwoh C K, Wong T Y and Liu J 2014 A survey on computer aided diagnosis for ocular diseases *BMC Med. Inf. Decis. Making* **14** 1–29

[3] Moccia S, De Momi E, El Hadji S and Mattos L S 2018 Blood vessel segmentation algorithms—review of methods, datasets and evaluation metrics *Comput. Methods Programs Biomed.* **158** 71–91

[4] Kanse S S and Yadav D M 2019 Retinal fundus image for glaucoma detection: a review and study *J. Intell. Syst.* **28** 43–56

[5] Li T, Bo W, Hu C, Kang H, Liu H, Wang K and Fu H 2021 Applications of deep learning in fundus images: a review *Med. Image Anal.* **69** 101971

[6] Gulshan V *et al* 2016 Development and validation of a deep learning algorithm for detection of diabetic retinopathy in retinal fundus photographs *JAMA* **316** 2402–10

[7] Szegedy C, Vanhoucke V, Ioffe S, Shlens J and Wojna Z 2016 Rethinking the inception architecture for computer vision *Proc. IEEE Conf. on Computer Vision and Pattern Recognition* pp 2818–26

[8] Krause J, Gulshan V, Rahimy E, Karth P, Widner K, Corrado G S, Peng L and Webster D R 2018 Grader variability and the importance of reference standards for evaluating machine learning models for diabetic retinopathy *Ophthalmology* **125** 1264–72

[9] Gondal W M, Köhler J M, Grzeszick R, Fink G A and Hirsch M 2017 Weakly-supervised localization of diabetic retinopathy lesions in retinal fundus images *IEEE Int. Conf. on Image Processing* (Piscataway, NJ: IEEE) pp 2069–73

[10] Wang Z, Yin Y, Shi J, Fang W, Li H and Wang X 2017 Zoom-in-net: deep mining lesions for diabetic retinopathy detection *Int. Conf. on Medical Image Computing and Computer-Assisted Intervention* (Cham: Springer) pp 267–75

[11] Pour A M, Seyedarabi H, Jahromi S H and Javadzadeh A 2020 Automatic detection and monitoring of diabetic retinopathy using efficient convolutional neural networks and contrast limited adaptive histogram equalization *IEEE Access* **8** 136668–73

[12] He A, Li T, Li N, Wang K and Fu H 2020 CABNet: category attention block for imbalanced diabetic retinopathy grading *IEEE Trans. Med. Imaging* **40** 143–53

[13] Ju L, Wang X, Wang L, Mahapatra D, Zhao X, Harandi M, Drummond T, Liu T and Ge Z 2021 Improving medical image classification with label noise using dual-uncertainty estimation arXiv: 2103.00528

[14] Araújo T, Aresta G, Mendonça L, Penas S, Maia C, Carneiro Â, Mendonça A M and Campilho A 2020 DR|GRADUATE: uncertainty-aware deep learning-based diabetic retinopathy grading in eye fundus images *Med. Image Anal.* **63** 101715

[15] Shankar K, Zhang Y, Liu Y, Wu L and Chen C H 2020 Hyperparameter tuning deep learning for diabetic retinopathy fundus image classification *IEEE Access* **8** 118164–73

[16] Majumder S and Kehtarnavaz N 2021 Multitasking deep learning model for detection of five stages of diabetic retinopathy arXiv: 2103.04207

[17] Lin Z, Guo R, Wang Y, Wu B, Chen T, Wang W, Chen D Z and Wu J 2018 A framework for identifying diabetic retinopathy based on anti-noise detection and attention-based fusion *Int. Conf. on Medical Image Computing and Computer-Assisted Intervention* (Cham: Springer) pp 74–82

[18] Yang Y, Li T, Li W, Wu H, Fan W and Zhang W 2017 Lesion detection and grading of diabetic retinopathy via two-stages deep convolutional neural networks *Int. Conf. on Medical Image Computing and Computer-Assisted Intervention* (Cham: Springer) pp 533–40

[19] Foo A, Hsu W, Lee M L, Lim G and Wong T Y 2020 Multi-task learning for diabetic retinopathy grading and lesion segmentation *Proc. AAAI Conf. Artificial Intelligence* vol 34 pp 13267–72

[20] Wong W L, Su X, Li X, Cheung C M, Klein R, Cheng C Y and Wong T Y 2014 Global prevalence of age-related macular degeneration and disease burden projection for 2020 and 2040: a systematic review and meta-analysis *Lancet Global Health* **2** e106–16

[21] Bird A C *et al* 1995 An international classification and grading system for age-related maculopathy and age-related macular degeneration *Surv. Ophthalmol.* **39** 367–74

[22] Hageman G S and Mullins R F 1999 Molecular composition of drusen as related to substructural phenotype *Mol. Vis.* **5** 10562652

[23] Govindaiah A, Hussain M A, Smith R T and Bhuiyan A 2018 Deep convolutional neural network based screening and assessment of age-related macular degeneration from fundus images *IEEE 15th Int. Symp. on Biomedical Imaging* (Piscataway, NJ: IEEE) pp 1525–8

[24] Simonyan K and Zisserman A 2014 Very deep convolutional networks for large-scale image recognition arXiv:1409.1556

[25] Grassmann F, Mengelkamp J, Brandl C, Harsch S, Zimmermann M E, Linkohr B, Peters A, Heid I M, Palm C and Weber B H 2018 A deep learning algorithm for prediction of age-related eye disease study severity scale for age-related macular degeneration from color fundus photography *Ophthalmology* **125** 1410–20

[26] Peng Y, Dharssi S, Chen Q, Keenan T D, Agrón E, Wong W T, Chew E Y and Lu Z 2019 Deep see net: a deep learning model for automated classification of patient-based age-related macular degeneration severity from color fundus photographs *Ophthalmology* **126** 565–75

[27] Sapieha P, Joyal J S, Rivera J C, Kermorvant-Duchemin E, Sennlaub F, Hardy P, Lachapelle P and Chemtob S 2010 Retinopathy of prematurity: understanding ischemic retinal vasculopathies at an extreme of life *J. Clin. Investigation* **120** 3022–32

[28] Zhang R, Zhao J, Chen G, Wang T, Zhang G and Lei B 2019 Aggressive posterior retinopathy of prematurity automated diagnosis via a deep convolutional network *Int. Workshop on Ophthalmic Medical Image Analysis (17 Oct.)* (Cham: Springer) pp 165–72

[29] Peng Y, Zhu W, Chen F, Xiang D and Chen X 2020 Automated retinopathy of prematurity screening using deep neural network with attention mechanism *Proc. SPIE* **11313** 1131321

[30] Hu J, Chen Y, Zhong J, Ju R and Yi Z 2018 Automated analysis for retinopathy of prematurity by deep neural networks *IEEE Trans. Med. Imaging* **38** 269–79

[31] Rani P, Rajkumar E R, Rajamani K T, Kandemir M and Singh D 2015 Detection of retinopathy of prematurity using multiple instance learning *Int. Conf. Advances in Computing, Communications and Informatics* (Piscataway, NJ: IEEE) pp 2233–7

[32] Chen G, Zhao J, Zhang R, Wang T, Zhang G and Lei B 2019 Automated stage analysis of retinopathy of prematurity using joint segmentation and multi-instance learning *Int. Workshop on Ophthalmic Medical Image Analysis* (Cham: Springer) pp 173–81

[33] Li X *et al* 2021 Deep multiple instance learning with spatial attention for ROP case classification, instance selection and abnormality localization *25th Int. Conf. on Pattern Recognition* (Piscataway, NJ: IEEE) pp 7293–8
[34] Zhou Y, Li G and Li H 2019 Automatic cataract classification using deep neural network with discrete state transition *IEEE Trans. Med. Imaging* **39** 436–46
[35] Xu X, Zhang L, Li J, Guan Y and Zhang L 2019 A hybrid global-local representation CNN model for automatic cataract grading *IEEE J. Biomed. Health Inform.* **24** 556–67
[36] Quellec G, Lamard M, Conze P H, Massin P and Cochener B 2020 Automatic detection of rare pathologies in fundus photographs using few-shot learning *Med. Image Anal.* **61** 101660
[37] Ju L, Wang X, Wang L, Liu T, Zhao X, Drummond T, Mahapatra D and Ge Z 2021 Relational subsets knowledge distillation for long-tailed retinal diseases recognition arXiv: 2104.11057
[38] Pachade S, Porwal P, Thulkar D, Kokare M, Deshmukh G, Sahasrabuddhe V, Giancardo L, Quellec G and Mériaudeau F 2021 Retinal fundus multi-disease image dataset (RFMiD): a dataset for multi-disease detection research *Data* **6** 14
[39] Li X, Hu X, Yu L, Zhu L, Fu C W and Heng P A 2019 CAnet: cross-disease attention network for joint diabetic retinopathy and diabetic macular edema grading *IEEE Trans. Med. Imaging* **39** 1483–93
[40] Tu Z, Gao S, Zhou K, Chen X, Fu H, Gu Z, Cheng J, Yu Z and Liu J 2020 SUNet: A lesion regularized model for simultaneous diabetic retinopathy and diabetic macular edema grading *IEEE 17th Int. Symp. on Biomedical Imaging* (Piscataway, NJ: IEEE) pp 1378–82
[41] Ju L, Wang X, Zhao X, Lu H, Mahapatra D, Bonnington P and Ge Z 2021 Synergic adversarial abel learning for grading retinal diseases via knowledge distillation and multi-task learning *IEEE J. Biomed. Health Informatics* **25** 3709–20
[42] Ting D S *et al* 2017 Development and validation of a deep learning system for diabetic retinopathy and related eye diseases using retinal images from multiethnic populations with diabetes *JAMA* **318** 2211–23
[43] Li C, Ye J, He J, Wang S, Qiao Y and Gu L 2020 Dense correlation network for automated multi-label ocular disease detection with paired color fundus photographs *17th Int. Symp. on Biomedical Imaging* (Piscataway, NJ: IEEE) pp 1–4
[44] Wang X, Ju L, Zhao X and Ge Z 2019 Retinal abnormalities recognition using regional multitask learning *Int. Conf. on Medical Image Computing and Computer-Assisted Intervention* (Cham: Springer) pp 30–8
[45] Redmon J and Farhadi A 2018 YOLOv3: an incremental improvement arXiv: 1804.02767
[46] Wang J, Yang Y, Mao J, Huang Z, Huang C and Xu W 2016 CNN-RNN: a unified framework for multi-label image classification *Proc. IEEE Conf. on Computer Vision and Pattern Recognition* pp 2285–94
[47] Decenciere E *et al* 2013 TeleOphta: machine learning and image processing methods for teleophthalmology *IRBM* **34** 196–203
[48] Porwal P, Pachade S, Kamble R, Kokare M, Deshmukh G, Sahasrabuddhe V and Meriaudeau F 2018 Indian diabetic retinopathy image dataset (IDRiD): a database for diabetic retinopathy screening research *Data* **3** 25
[49] Li T, Gao Y, Wang K, Guo S, Liu H and Kang H 2019 Diagnostic assessment of deep learning algorithms for diabetic retinopathy screening *Inf. Sci.* **501** 511–22

[50] Kauppi T, Kalesnykiene V, Kamarainen J K, Lensu L, Sorri I, Raninen A, Voutilainen R, Uusitalo H, Kälviäinen H and Pietilä J 2007 The DIARETDB1 diabetic retinopathy database and evaluation protocol *Proc. British Machine Vision Conf.* **1** 1–10

[51] Guo S, Li T, Kang H, Li N, Zhang Y and Wang K 2019 L-Seg: an end-to-end unified framework for multi-lesion segmentation of fundus images *Neurocomputing* **349** 52–63

[52] Xue J, Yan S, Qu J, Qi F, Qiu C, Zhang H, Chen M, Liu T, Li D and Liu X 2019 Deep membrane systems for multi-task segmentation in diabetic retinopathy *Knowl.-Based Syst.* **183** 104887

[53] Baudoin C E, Lay B J and Klein J C 1984 Automatic detection of microaneurysms in diabetic fluorescein angiography *Rev. Epidémiol. Santé Publique* **32** 254–61

[54] Walter T, Massin P, Erginay A, Ordonez R, Jeulin C and Klein J C 2007 Automatic detection of microaneurysms in color fundus images *Med. Image Anal.* **11** 555–66

[55] He K, Gkioxari G, Dollár P and Girshick R 2017 Mask R-CNN *Proc. IEEE Int. Conf. on Computer Vision* pp 2961–9

[56] Wang X, Xu M, Zhang J, Jiang L and Li L 2021 Deep multi-task learning for diabetic retinopathy grading in fundus images *Proc. AAAI Conf. on Artificial Intelligence* vol 35 pp 2826–34

[57] Zhou C, Zhang T, Wen Y, Chen L, Zhang L and Chen J 2021 Cross-modal guidance for hyperfluorescence segmentation in fundus fluorescein angiography *IEEE Int. Conf. on Multimedia and Expo* (Piscataway, NJ: IEEE) pp 1–6

[58] Xu Y, Zhou Z, Li X, Zhang N, Zhang M and Wei P 2021 FFU-Net: feature fusion U-Net for lesion segmentation of diabetic retinopathy *BioMed. Res. Int.* **2021** 6644071

[59] Zhao Y, Xie J, Su P, Zheng Y, Liu Y, Cheng J and Liu J 2018 Retinal artery and vein classification via dominant sets clustering-based vascular topology estimation *Int. Conf. on Medical Image Computing and Computer-Assisted Intervention* (Cham: Springer) pp 56–64

[60] Zheng H, Chang L, Wei T, Qiu X, Lin P and Wang Y 2016 Registering retinal vessel images from local to global via multiscale and multicycle features *Proc. IEEE Conf. on Computer Vision and Pattern Recognition Workshops* pp 50–7

[61] Waheed Z, Akram M U, Waheed A, Khan M A, Shaukat A and Ishaq M 2016 Person identification using vascular and non-vascular retinal features *Comput. Electr. Eng.* **53** 359–71

[62] Saker S 2016 Diabetic retinopathy: *in vitro* and clinical studies of mechanisms and pharmacological treatments *Doctoral Dissertation* University of Nottingham

[63] Niemeijer M, Staal J, van Ginneken B, Loog M and Abramoff M D 2004 Comparative study of retinal vessel segmentation methods on a new publicly available database *Proc. SPIE* **5370** 648–56

[64] Hoover A D, Kouznetsova V and Goldbaum M 2000 Locating blood vessels in retinal images by piecewise threshold probing of a matched filter response *IEEE Trans. Med. Imaging* **19** 203–10

[65] Fraz M M, Remagnino P, Hoppe A, Uyyanonvara B, Rudnicka A R, Owen C G and Barman S A 2012 An ensemble classification-based approach applied to retinal blood vessel segmentation *IEEE Trans. Biomed. Eng.* **59** 2538–48

[66] Budai A, Bock R, Maier A, Hornegger J and Michelson G 2013 Robust vessel segmentation in fundus images *Int. J. Biomed. Imaging* **2013** 154860

[67] Soares J V, Leandro J J, Cesar R M, Jelinek H F and Cree M J 2006 Retinal vessel segmentation using the 2-D Gabor wavelet and supervised classification *IEEE Trans. Med. Imaging* **25** 1214–22

[68] Yan Z, Yang X and Cheng K T 2018 A three-stage deep learning model for accurate retinal vessel segmentation *IEEE J. Biomed. Health Inform.* **23** 1427–36
[69] Wang D, Haytham A, Pottenburgh J, Saeedi O and Tao Y 2020 Hard attention net for automatic retinal vessel segmentation *IEEE J. Biomed. Health Inform.* **24** 3384–96
[70] Zhang S, Fu H, Xu Y, Liu Y and Tan M 2020 Retinal image segmentation with a structure-texture demixing network *Int. Conf. on Medical Image Computing and Computer-Assisted Intervention* (Cham: Springer) pp 765–74
[71] Wu Y, Xia Y, Song Y, Zhang D, Liu D, Zhang C and Cai W 2019 Vessel-Net: retinal vessel segmentation under multi-path supervision *Int. Conf. on Medical Image Computing and Computer-Assisted Intervention* (Cham: Springer) pp 264–72
[72] Wang B, Qiu S and He H 2019 Dual encoding U-Net for retinal vessel segmentation *Int. Conf. on Medical Image Computing and Computer-Assisted Intervention* (Cham: Springer) pp 84–92
[73] Xu R, Liu T, Ye X, Lin L and Chen Y W 2020 Boosting connectivity in retinal vessel segmentation via a recursive semantics-guided network *Int. Conf. on Medical Image Computing and Computer-Assisted Intervention* (Cham: Springer) pp 786–95
[74] Wang W, Zhong J, Wu H, Wen Z and Qin J 2020 RVSeg-Net: an efficient feature pyramid cascade network for retinal vessel segmentation *Int. Conf. on Medical Image Computing and Computer-Assisted Intervention* (Cham: Springer) pp 796–805
[75] Wu H, Wang W, Zhong J, Lei B, Wen Z and Qin J 2021 SCS-Net: a scale and context sensitive network for retinal vessel segmentation *Med. Image Anal.* **70** 102025
[76] Li K, Qi X, Luo Y, Yao Z, Zhou X and Sun M 2020 Accurate retinal vessel segmentation in color fundus images via fully attention-based networks *IEEE J. Biomed. Health Inform.* **25** 2071–81
[77] Xu R, Liu T, Ye X, Liu F, Lin L, Li L, Tanaka S and Chen Y W 2020 Joint extraction of retinal vessels and centerlines based on deep semantics and multi-scaled cross-task aggregation *IEEE J. Biomed. Health Inform.* **25** 2722–32
[78] Mou L, Chen L, Cheng J, Gu Z, Zhao Y and Liu J 2019 Dense dilated network with probability regularized walk for vessel detection *IEEE Trans. Med. Imaging* **39** 1392–403
[79] Zhang M, Yu F, Zhao J, Zhang L and Li Q 2020 BEFD: boundary enhancement and feature denoising for vessel segmentation *Int. Conf. on Medical Image Computing and Computer-Assisted Intervention* (Cham: Springer) pp 775–85
[80] Mary V S, Rajsingh E B and Naik G R 2016 Retinal fundus image analysis for diagnosis of glaucoma: a comprehensive survey *IEEE Access* **4** 4327–54
[81] Naithani P, Sihota R, Sony P, Dada T, Gupta V, Kondal D and Pandey R M 2007 Evaluation of optical coherence tomography and Heidelberg retinal tomography parameters in detecting early and moderate glaucoma *Investigative Ophthalmol. Vis. Sci.* **48** 3138–45
[82] Maninis K K, Pont-Tuset J, Arbeláez P and Van Gool L 2016 Deep retinal image understanding *Int. Conf. on Medical Image Computing and Computer-Assisted Intervention* (Cham: Springer) pp 140–8
[83] Lowell J, Hunter A, Steel D, Basu A, Ryder R, Fletcher E and Kennedy L 2004 Optic nerve head segmentation *IEEE Trans. Med. Imaging* **23** 256–64
[84] Aquino A, Gegúndez-Arias M E and Marín D 2010 Detecting the optic disc boundary in digital fundus images using morphological, edge detection, and feature extraction techniques *IEEE Trans. Med. Imaging* **29** 1860–9
[85] Cheng J, Liu J, Xu Y, Yin F, Wong D W, Tan N M, Tao D, Cheng C Y, Aung T and Wong T Y 2013 Superpixel classification based optic disc and optic cup segmentation for glaucoma screening *IEEE Trans. Med. Imaging* **32** 1019–32

[86] Wong D W, Liu J, Lim J H, Jia X, Yin F, Li H and Wong T Y 2008 Level-set based automatic cup-to-disc ratio determination using retinal fundus images in ARGALI *30th Annual Int. Conf. of the IEEE Engineering in Medicine and Biology Society* (Piscataway, NJ: IEEE) pp 2266–9

[87] Mohan D, Kumar J H and Seelamantula C S 2019 Optic disc segmentation using cascaded multiresolution convolutional neural networks *IEEE Int. Conf. on Image Processing* (Piscataway, NJ: IEEE) pp 834–8

[88] Gu Z, Cheng J, Fu H, Zhou K, Hao H, Zhao Y, Zhang T, Gao S and Liu J 2019 CE-Net: Context encoder network for 2D medical image segmentation *IEEE Trans. Med. Imaging* **38** 2281–92

[89] Al-Bander B, Williams B M, Al-Nuaimy W, Al-Taee M A, Pratt H and Zheng Y 2018 Dense fully convolutional segmentation of the optic disc and cup in colour fundus for glaucoma diagnosis *Symmetry* **10** 87

[90] Shankaranarayana S M, Ram K, Mitra K and Sivaprakasam M 2017 Joint optic disc and cup segmentation using fully convolutional and adversarial networks *Fetal, Infant and Ophthalmic Medical Image Analysis* (Cham: Springer) pp 168–76

[91] Jiang Y, Tan N and Peng T 2019 Optic disc and cup segmentation based on deep convolutional generative adversarial networks *IEEE Access* **7** 64483–93

[92] Yu S, Xiao D, Frost S and Kanagasingam Y 20191 Robust optic disc and cup segmentation with deep learning for glaucoma detection *Comput. Med. Imaging Graph.* **74** 61–71

[93] Zhang Z, Fu H, Dai H, Shen J, Pang Y and Shao L 2019 ET-Net: a generic edge-attention guidance network for medical image segmentation *Int. Conf. on Medical Image Computing and Computer-Assisted Intervention* (Cham: Springer) pp 442–50

[94] Xu Y L, Lu S, Li H X and Li R R 2019 Mixed maximum loss design for optic disc and optic cup segmentation with deep learning from imbalanced samples *Sensors* **19** 4401

[95] Wang Z, Dong N, Rosario S D, Xu M, Xie P and Xing E P 2019 Ellipse detection of optic disc-and-cup boundary in fundus images *IEEE 16th Int. Symp. on Biomedical Imaging* (Piscataway, NJ: IEEE) pp 601–4

[96] Jiang Y, Duan L, Cheng J, Gu Z, Xia H, Fu H, Li C and Liu J 2019 JointRCNN: a region-based convolutional neural network for optic disc and cup segmentation *IEEE Trans. Biomed. Eng.* **67** 335–43

[97] Murugesan B, Sarveswaran K, Shankaranarayana S M, Ram K, Joseph J and Sivaprakasam M 2019 Psi-Net: shape and boundary aware joint multi-task deep network for medical image segmentation *41st Annual Int. Conf. of the IEEE Engineering in Medicine and Biology Society* (Piscataway, NJ: IEEE) pp 7223–6

[98] Fu H, Cheng J, Xu Y, Wong D W, Liu J and Cao X 2018 Joint optic disc and cup segmentation based on multi-label deep network and polar transformation *IEEE Trans. Med. Imaging* **37** 1597–605

[99] Liu Q, Hong X, Li S, Chen Z, Zhao G and Zou B 2019 A spatial-aware joint optic disc and cup segmentation method *Neurocomputing* **359** 285–97

[100] Edupuganti V G, Chawla A and Kale A 2018 Automatic optic disk and cup segmentation of fundus images using deep learning *25th IEEE Int. Conf. on Image Processing* (Piscataway, NJ: IEEE) pp 2227–31

[101] Mohan D, Kumar J H and Seelamantula C S 2018 High-performance optic disc segmentation using convolutional neural networks *25th IEEE Int. Conf. on Image Processing* (Piscataway, NJ: IEEE) pp 4038–42

[102] Shah S, Kasukurthi N and Pande H 2019 Dynamic region proposal networks for semantic segmentation in automated glaucoma screening *IEEE 16th Int. Symp. on Biomedical Imaging* (Piscataway, NJ: IEEE) pp 578–82

[103] Wang S, Yu L, Yang X, Fu C W and Heng P A 2019 Patch-based output space adversarial learning for joint optic disc and cup segmentation *IEEE Trans. Med. Imaging* **38** 2485–95

[104] Wang S, Yu L, Li K, Yang X, Fu C W and Heng P A 2019 Boundary and entropy-driven adversarial learning for fundus image segmentation *Int. Conf. on Medical Image Computing and Computer-Assisted Intervention* (Cham: Springer) pp 102–10

[105] Yin P, Wu Q, Xu Y, Min H, Yang M, Zhang Y and Tan M 2019 PM-Net: pyramid multi-label network for joint optic disc and cup segmentation *Int. Conf. on Medical Image Computing and Computer-Assisted Intervention* (Cham: Springer) pp 129–37

[106] Liu P, Kong B, Li Z, Zhang S and Fang R 2019 CFEA: collaborative feature ensembling adaptation for domain adaptation in unsupervised optic disc and cup segmentation *Int. Conf. on Medical Image Computing and Computer-Assisted Intervention* (Cham: Springer) pp 521–9

[107] Zhang Z, Yin F S, Liu J, Wong W K, Tan N M, Lee B H, Cheng J and Wong T Y 2010 ORIGA-light: an online retinal fundus image database for glaucoma analysis and research *Annual Int. Conf. of the IEEE Engineering in Medicine and Biology* (Piscataway, NJ: IEEE) pp 3065–8

[108] Fumero F, Alayón S, Sanchez J L, Sigut J and Gonzalez-Hernandez M 2011 RIM-ONE: an open retinal image database for optic nerve evaluation *24th Int. Symp. on Computer-based Medical Systems* (Piscataway, NJ: IEEE) pp 1–6

[109] Sivaswamy J, Krishnadas S R, Joshi G D, Jain M and Tabish A U 2014 Drishti-GS: retinal image dataset for optic nerve head (ONH) segmentation *IEEE 11th Int. Symp. on Biomedical Imaging* (Piscataway, NJ: IEEE) pp 53–6

IOP Publishing

Photo Acoustic and Optical Coherence Tomography Imaging, Volume 2

Fundus imaging for the retina

Ayman El-Baz and Jasjit S Suri

Chapter 9

Diabetic retinopathy detection and classification through fundus images using AlexNet

Amnia Salma

9.1 Introduction

The number of diabetes mellitus (DM) patients has increased from 180 million in 1980 to 422 million in 2020 [1]. The data show that DM has become a threat for people around the world. One complication of DM is diabetic retinopathy (DR). At a severe level, DR can cause blindness [2, 3]. An early sign of DR is microaneurysms. A microaneurysm is seen as a little red spot in the retina [4, 5]. The other sign of DR is hemorrhages or red lesions because of broken vessels and hard exudates [6]. Neovascularization is one of the signs of DR at a severe level [7]. DR is divided into three levels: normal or no DR, non-proliferative diabetic retinopathy (NPDR), and proliferative diabetic retinopathy (PDR). Patient screening is used to detect DR, but the number of ophthalmologists is small and does not compare with the number of DR patients. The examination takes too long a time, is very costly, and needs professional skills to perform.

Many researchers have developed image classification systems to detect and classify disease. One of them is DR detection and classification using fundus images [8–12]. Computer vision, machine learning, and deep learning can be used to detect some diseases by using image classification for medical health [13]. DR diagnosis is performed through feature extraction from fundus images with machine learning. In the big data era, many developments of computer vision using deep learning show very good accuracy for DR detection [14]. Many researchers use the convolutional neural network (CNN) model for image classification, such as DR detection. One of the architectures of CNN is AlexNet, which has been used for DR classification with high accuracy [15, 16].

Technological developments in the field of artificial intelligence (AI) can be used to detect several diseases such as DR automatically with accurate results through fundus image data [6]. The development of AI includes machine learning (ML), which is a system that includes the design and development of algorithms that allow computers to develop behavior based on empirical data, such as from sensor data and databases.

Based on research led by the authors of [17], the diagnosis of DR can be performed through feature extraction and feature selection on fundus images using machine learning, namely the support vector machine (SVM) and *K*-nearest neighbors (KNN) methods which obtain 80%–84% accuracy for the classification of three classes, namely normal, DR, and age-related macular degradation (AMD), and 100% for the classification of two classes, namely normal and DR. Feature textures in fundus images are easier to implement than lesion characteristics [18]. Sarwinda, Alhadi, and Titin (2018) utilized machine learning in the extraction process in fundus images at three levels of severity of DR, namely normal, mild NPDR, and moderate/severe NPDR. However, the use of machine learning in the DR classification still has several limitations, including the use of image data ranging from 125 to 154 samples, which is considered small [10].

In deep learning, one CNN architecture [15] is the large deep CNN known as AlexNet, used to classify images. AlexNet supports the processing of large amounts of image data and with fast image processing speed compared to other architectures. AlexNet also received a top-5 error predicate of 16%. Several studies utilizing CNN were conducted by [20] using the CNN method to classify DR disease into five classes, namely normal, mild, moderate, severe, and proliferative DR. The authors of [14] used the CNN method to validate and develop DL for DR with retinal images from a multiethnic group with a history of diabetes, the accuracy of which was more than 90%. The authors of [20] classified the level of DR disease through fundus images using transfer learning and hyperparameter tuning methods by comparing all CNN architectures and obtaining 95.68% accuracy results.

In this study we describe how to use the AlexNet architecture to classify fundus image to detect DR. The output of this research classifies DR into three classes: normal or no DR, NPDR, and PDR. This classification is used to help ophthalmologists detect DR. Early detection can help patients receive suitable treatment and prevent blindness.

9.2 Diabetic retinopathy detection and classification

9.2.1 Diabetic retinopathy

Diabetic retinopathy (DR) is a complication of diabetes caused by high sugar levels causing damage to the blood vessels in the retina, particularly in light-sensitive tissues. Data compiled by TheDiabCare Asia in 2014 showed that 42% of people with diabetes mellitus in Indonesia experienced complications in the form of DR and 6.4% were at the proliferative level of DR.

Medically, diabetic retinopathy is generally divided into two classes based on the presence or absence of new blood vessels in the retina: non-proliferative diabetic retinopathy (NPDR) and proliferative diabetic retinopathy (PDR). This study divides DR into three levels: normal, NPDR, and PDR.

The earliest sign of DR disease is microaneurysms, seen as small red dots on the retina. Another sign is hemorrhages or bleeding, which are observed as red lesions caused by the rupture of small blood vessels in the deeper layers of the retina. In addition, there are also signs of exudate, a whitish-yellow lesion caused by capillary plasma leakage [21].

According to [22], diabetes patients in the early stages of DR show almost no symptoms at all, but over time symptoms can appear and occur in one or both eyes. Signs of DR are:

(i) Vision decreases gradually.
(ii) Black spots occur when viewing.
(iii) Floaters or stains that float in the field of vision.
(iv) Vision becomes shaded.
(v) Colors become difficult to distinguish.
(vi) Eyes feel sore or appear red.
(vii) Loss of vision (blindness).

9.2.2 Convolutional neural network

The convolutional neural network (CNN) is a part of deep learning which has a variety of architectures that are useful for classifying images, text, sound, video, and so on. There are several types of layers in CNNs, such as convolutional layers, pooling layers, and fully connected layers.

A convolutional layer is the first layer in a CNN which is useful for feature extraction, which is then assisted by the next process in the form of the rectified linear unit (ReLu), which is useful for deactivating unused image pixels and only maintaining pixels that are considered important. The pooling layer is useful for reducing image size so that the computing process runs faster and does not take up more storage space. The last layer in the CNN is the fully connected layer, where the layer takes input from the number of values derived from the previous layers and provides information/data to the neural network. The output layer is the result of a series of computational processes. In the context of image classification, the number of existing outputs is an interpretation of the output results that have been set and adjusted to the number of classes of the data entered.

Basically, an image is seen as a collection of numbers in a matrix, where each number has a sign of the light intensity at that point, commonly called a pixel. An image has three parameters that need to be considered, namely height, width, and depth, which can be written as $h \times w \times d$.

In its most general form, convolution is a real-valued operation of two-argument functions. In mathematical functions, convolution is the operation of two functions into w, which results in the function where $(x*w)$, which tells how one form is modified by the other. Convolution refers to the output function and the result of its calculation. Convolution can be expressed by the equation

$$s(t) = \int x(a)w(t - a)\mathrm{d}a,$$

$$s(t) = (x*w)(t).$$

The function in the first argument is called the input, and the function in the second argument is called a filter; (t) is the result of a convolution which is commonly known as a feature map.

The use of discrete data, t, has an integer index, so x and y generate only part of the integer contained in t. The convolution equation for discrete data is as follows:

$$s(t) = (x*w)(t) = \sum\nolimits_{a=-\infty}^{\infty} x(a)w(t - a).$$

This study uses an input (I) image in the form of 2D, which has more than one axis, so the filter (F) used is in the form of 2D. The form of the equation is as follows:

$$s(i, j) = (F*I)(i, j) = \sum_{m}\sum_{n} I(m, n)F(i - m, j - n).$$

By applying the commutative property according to convolution, the equation above can be written as

$$s(i, j) = (F*I)(i, j) = \sum_{m}\sum_{n} I(i - m, j - n)F(m, n),$$

where the sign (*) indicates the convolution operator F, I indicates the kernel or filter of image input, s shows the output of the convolution result, i, j indicate the row index of the feature map column, and m, n represent the index of row and column length in the convolution layer filter.

9.2.2.1 Convolutional layer

The convolutional layer is the most central part of the CNN [23]. The process that occurs in the convolutional layer is in the form of feature extraction, where all the input parts are taken in small portions and passed through all the points in the initial data. After the data passes through the various processes it is combined into the output.

In the convolutional layer, two components are used, namely the input data matrix and the matrix with a value for extracting the input matrix, called a filter. Filters have a certain height, width, and thickness. The filter value becomes a parameter that will continue to be updated during the learning process. In this process, there is a multiplication between the matrix and the input value of image data, known as convolution. There are essential processes that are carried out before the convolution, between the input and convolutional layers, namely stride and padding. Stride is the value used as a reference for filter shifting. Padding is a reference value used to determine the number of pixels to be added to the input data.

9.2.2.2 Pooling layer

The pooling layer is the stage after the convolutional layer. The function of the pooling layer is to reduce the input spatially by a down-sampling operation. This is done to minimize the variance and complexity in computing so the computational process is faster and does not take a lot of memory. The pooling layer helps to control overfitting and progressively reduces data representation over the network. Some of the pooling techniques are max pooling, mean pooling, and average pooling.

In this study, the pooling technique used is max pooling, which is shown in figure 9.1.

Mathematically, the calculation of the output dimensions contained in the pooling layer is

$$O = \frac{I - K}{S} + 1,$$

where O represents the dimension of output pooling, I represents dimension of the input image, K represents the dimension matrix filter, and S represents the number of strides.

9.2.2.3 Fully connected layer

The fully connected layer is a layer that comes from all previous layer networks and is connected to networks in the next layer. Connections with all neurons in the last layer can be achieved by converting the previous layer into one-dimensional data. This layer is the last layer in the CNN after the convolutional layer and pooling layer which makes each layer connected so that it can carry out the classification process.

The fully connected layer is used for the data classification process. The following equation is the definition of a fully connected layer:

$$z_i = \sum_{j=1}^{c} w^T{}_{i,j} \mathrm{x}_i + b_{j,}$$

where i represents the number of features *input*, j represents the number of target class, z_i represents the *output* value of the neuron from feature i, x_i represents the *input* value from feature extraction in i, $w_{i,j}$ represents the weight from neuron size i x j, and b_j represents the error from the neuron of j.

The fully connected layer is always followed by a loss function. In this study we use the cross entropy method to calculate the loss value of the model. The loss function represents the error value generated by the model and is used when training provides errors that must be considered. In addition, the loss function is considered good if the expected function produces the lowest loss value. Thus, the evaluation of

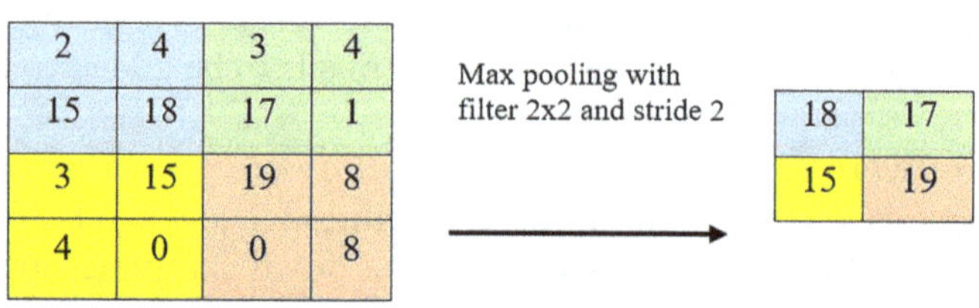

Figure 9.1. Illustration of the max pooling operation.

the model's performance is based on torch contributors (2019) with the loss criteria defined as follows:

$$\text{loss}\,(x, \text{class}) = -\log\left(\frac{\exp(x[\text{class}])}{\sum_j \exp(x[j])}\right)$$

$$\text{loss}\,(x, \text{class}) = -x[\text{class}] + \log\left(\sum_j \exp(x[j])\right)$$

$$\text{loss}\,(x, \text{class}) = -\text{weight}[\text{class}]\left(-x[\text{class}] + \log\left(\sum_j \exp\,(x[j])\right)\right)$$

$$\text{loss} = \frac{\sum_{i=1}^{N} \text{loss}(i, \text{class}[i])}{\sum_{i=1}^{N} \text{weight}(\text{class}[i])}.$$

9.2.3 Method

In recent years, the development of applied artificial intelligence for medical uses has been challenging for researchers. In the big data era, many methods are emerging for classifying images for disease detection. One of them is the CNN architecture, such as AlexNet. In this research we used the AlexNet architecture to classify fundus images into normal, NPDR, and PDR levels based on the sign shown in the fundus images.

9.2.3.1 Obtaining dataset

Obtaining the dataset is one of the challenging tasks. There are many datasets in online databases. The Kaggle dataset has many fundus images with high resolution. The images in the Kaggle dataset are indexed by a pathologist. We used a dataset from Kaggle for DR detection using image classification through fundus images. We used 197 fundus images, with 38 normal, 114 NPDR, and 45 PDR.

9.2.3.2 Preprocessing image

Preprocessing images form a considerable part of a CNN. The size of each image is 3152×3000 and the images are in .jpg format. The images are resized to fit the ALexNet architecture properly. The proper image size for AlexNet is 227×227 pixels.

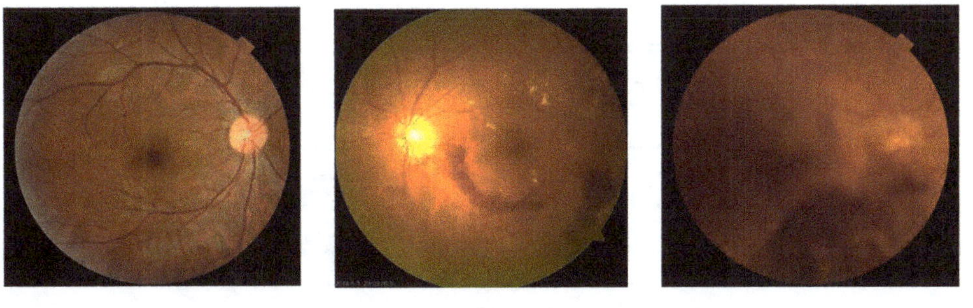

Figure 9.2. Fundus images: normal, NPDR, and PDR.

The color type of the images is RGB. Although RGB images are complex to process and require a lot of time, more features can be extracted from such images.

9.2.3.3 Training data

In this study, the AlexNet architecture used consists of five convolutional layers and three fully connected layers for the classification of three classes, namely normal, NPDR, and PDR. The layers consist of a convolutional layer, pooling layer, ReLu, activation function, and fully connected layer. The structure of the AlexNet architecture can be seen in figure 9.3.

The AlexNet architecture above has 96 hidden layers in the first convolution, 256 hidden layers in the second convolution, and 384 hidden layers in the third convolution. The fourth and fifth convolutions have 256 hidden layers. In the fully connected layer, there are 4096 hidden layers in the first fully connected and the second fully connected layers, and there is an ReLu activation function which is then taken to the last fully connected layer for classification.

9.2.3.4 Performance evaluation

The process of evaluating the performance of the model used in this study is based on accuracy and computation time. The accuracy of the model classification is calculated using the confusion matrix [24] (table 9.1).

Information:

(i) True positives (TPs) are the result where the model correctly predict the positive class.
(ii) False negatives (FNs) are the result where the model correctly predict the negative class.
(iii) False positives (FPs) are the result where the model incorrectly predict the positive class.
(iv) True negatives (TNs) are the result where the model incorrectly predict the negative class.

Performance evaluation of the classification model is carried out by calculating the level of accuracy, precision, recall, and F-1 score for each class of DR severity. Each is calculated using the following equation:

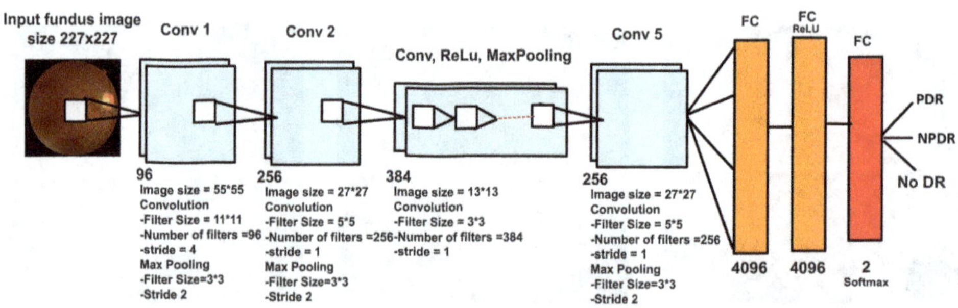

Figure 9.3. AlexNet architecture.

Table 9.1. Confusion matrix.

		Prediction		
		f_1	f_2	f_n
Actual	f_1	TF (f_1)	FP (f_1, f_2)	FP (f_1, f_n)
	f_2	FP (f_1, f_2)	TF (f_2)	FP (f_2, f_n)
	f_n	FP (f_1, f_n)	FP (f_2, f_n)	TN (f_n)

$$\text{Accuracy} = \frac{\text{TP} + \text{TN}}{\text{TP} + \text{FN} + \text{TN} + \text{TP}}$$

$$\text{Precision } (P) = \frac{\text{TP}}{\text{TP} + \text{FP}}$$

$$\text{Recall } (R) = \frac{\text{TN}}{\text{TN} + \text{FP}}$$

$$\text{F-1} = \frac{2RP}{R + P}.$$

The F-1 score is a balanced measure that can be used if the class has various sizes. The best value of the F-1 score is 1, and the lowest value is 0. A value of the F-1 score of 1 indicates that the model constructed has good precision and recall.

9.3 Results

9.3.1 AlexNet architecture

We used the AlexNet architecture for this model. First, the neural network trained on 80% of the dataset, and then another 20% were used for testing. For image processing, the image was resized to fit AlexNet, at 227 × 227 pixels. Every image was divided into multiple pieces of 227 × 227 pixels. Every piece was given a specific weight, where the area in the center of the eye has the highest weight. The value of a higher weight represents greater intensity of the disease. Thus the fundus, which has a high weight, is indicated in the DR category. We trained the model using Python programming and the Pytorch library. Every piece of of the multiple pieces was input to the model. Our model had 96 hidden layers in the first convolutional layer and 256 hidden layers in the second convolutional layer. In the third convolutional layer we had 384 hidden layers, and then in the fourth and fifth convolutional layers we had 256 hidden layers—every convolutional layer followed by max pooling and activation functions. In the fully connected layer we had 4096 hidden layers and the second fully connected layer was followed by ReLu, and the last fully connected layer was followed by softmax. We set the output of softmax by 3, which indicates three output classifications (table 9.2).

In the AlexNet architecture, the architecture-compliant image input is 227 × 227 pixels. The training process begins with a convolutional layer process which aims to

Table 9.2. Results for the model of AlexNet. Values are in pixels.

Layer name	Input size	Filter size/stride/padd.	Pooling size/stride/padd.	Output
Conv.	227 × 227 × 3	11 × 11 × 96/4/0	—	55 × 55 × 96
Max pool	55 × 55 × 94	—	3 × 3/2/0	27 × 27 × 96
Conv.	27 × 27 × 96	5 × 5 × 256/1	—	27 × 27 × 256
Max pool.	27 × 27 × 256	—	3 × 3/2/0	13 × 13 × 256
Conv.	13 × 13 × 256	3 × 3 × 384/1/1	—	13 × 13 × 384
Conv.	13 × 13 × 384	3 × 3 × 384/1/1		13 × 13 × 256
Conv.	13 × 13 × 384	3 × 3 × 256/1/1	—	6 × 6 × 256
Max pool	13 × 13 × 256	—	3 × 3/2/0	4096
Fully connected	—	—	—	4096
Softmax	—	—	—	3

Table 9.3. Confusion matrix of AlexNet.

		Predicted		
		Normal	NPDR	PDR
Actual	Normal	12	0	1
	NPDR	1	12	0
	PDR	0	0	14

multiply the matrix with a filter size value of 11 × 11 × 96 and stride 4 and padding 0 pixels. This process produces an output value of 55 × 55 × 96 pixels. After that, the pooling layer is entered with input size 55 × 55 × 96, filter size 5 × 5 × 256, and stride 1 pixels. The process is repeated twice and enters the convolutional layer three times and the pooling layer once. After that, they data enter the last layer in the form of a fully connected layer with softmax so as to produce five classes of output.

9.3.2 The result

The output of this model is analyzed. We used 400 epochs and the learning rate is 0.001. The model performance in terms of accuracy, precision, recall, and F-1 score can be seen in table 9.3. The performance of models achieves very good accuracy up to 94%.

Further analysis was carried out using the confusion matrix method to determine the model's performance in each class. The confusion matrix for the model with AlexNet is shown in table 9.4.

Based on the results of the data tested using the AlexNet architecture, it can be concluded that the accuracy achieved is as described in table 9.4, with an overall accuracy of 94% with the highest accuracy details in the PDR 100% class, as well as the best precision, recall, and F-1 scores. The NPDR class had values of 100%, 92%,

Table 9.4. Performance accuracy, precision, recall, and F-1 score.

Class	Accuracy	Precision	Recall	F-1 score
Normal	0.92	0.92	0.92	**0.92**
NPDR	0.92	1	0.92	**0.96**
PDR	1	0.93	1	**0.97**

and 96%, respectively. In contrast, the lowest scores for each category of accuracy, precision, recall, and F-1 score are in the normal class with values of 92%. In general, all the accuracy values obtained are very good, above 90%. This happens because the model classifies images properly, particularly in the PDR class.

9.4 Conclusion

Our model, using the AlexNet architecture, trained 197 fundus images from the Kaggle dataset to classify three classes: normal, NPDR, and PDR. Before training the data, we used image preprocessing where the images were resized to fit AlexNet input image requirement. By this process, we know the weight of each input. After that, we train the data with the AlexNet architecture to see the classification result. This study represents a new model for the detection of diabetic retinopathy disease. Using the AlexNet CNN architecture, the result gives an accuracy of 94%. The best class based on the confusion matrix is PDR with accuracy of 100%, precision of 93%, recall of 100%, and F-1 score of 97%.

In this paper, we use the model to implement deep learning from fundus images. In addition, this system was created to help ophthalmologists detect diabetic retinopathy, not to replace doctors. Furthermore, it can also perform multiclass classification for diabetic retinopathy.

References

[1] WHO 2019 Diabetes https://www.who.int/news-room/fact-sheets/detail/diabetes (Accessed: 21 December)

[2] Mookiah M R K, Acharya U R, Chua C K, Lim C M, Ng E Y K and Laude A 2013 Computer-aided diagnosis of diabetic retinopathy: a review *Comput. Biol. Med.* **43** 2136–55

[3] Saeed E, Szimkowsky M, Saeed K and Mariak Z 2019 An approach to automatic hard exudate detection in retina color images by a telemedicine system based on the d-eye sensor and image processing algorithms *Sensors* **19** 695

[4] Indumathi G and Sathananthavathi V 2019 Microaneurysms detection for early diagnosis of diabetic retinopathy using shape and steerable Gaussian features *Telemedicine Technologies* (Amsterdam: Elsevier) pp 57–69

[5] Jaya T, Dheeba J and Singh N A 2015 Detection of hard exudates in colour fundus images using fuzzy support vector machine-based expert system *J. Digit. Imaging* **28** 761–8

[6] Kohjasteh P, Aliahmad B and Kumar D K 2018 Fundus images analysis using deep features for detection of exudates, hemorrhages and microaneurysms *BMC Ophthalmol.* **18** 275–88

[7] Mahiba C and Jayachandran A 2019 Severity analysis of diabetic retinopathy in retinal images using hybrid structure descriptor and modified CNNs *Measurement* **135** 762–7
[8] Salamat N, Missen M M S and Rashid A 2019 Diaetic retinopathy techniques in retinal images: a review *Artif. Intell. Med.* **97** 168–88
[9] Sangeetha S N and Uma P U 2018 An intelligent model for blood vessel segmentation in diagnosing DR using CNN *J. Med. Syst.* **42** 175–85
[10] Sarwinda D, Bustamam A and Arymurthy A M 2017 Fundus image texture features analysis in diabetic retinopathy diagnosis *Int. Conf. on Sensing Technology* (Piscataway, NJ: IEEE)
[11] Venkatesan R, Chandakkar P, Li B and Li H K 2012 Classification of diabetic retinopathy images using multi-class multiple-instance learning based on color correlogram features *Ann. Int. Conf. of the IEEE Engineering in Medicine and Biology Society* (Piscataway, NJ: IEEE) pp 1462–5
[12] Yi-Peng L, Li Z, Xu C, Li J and Liang R 2019 Referable diabetic retinopathy identification from eye fundus images with weighted path for convolutional neural network *Artif. Intell. Med.* **99** 101694
[13] Maier A, Syben C, Lasser T and Riess C 2019 A gentle introduction to deep learning in medical image processing *Z. Med. Phys.* **29** 86–101
[14] Ting D S W *et al* 2017 Development and validation of a deep learning system *JAMA* **318** 2211–23
[15] Khrizhevsky A, Sutskever I and Hinton G E 2012 ImageNet classification with deep convolutional neural networks *Commun. ACM* **60** 84–90
[16] Shanthi T and Sabeenian R S 2019 Modified AlexNet architecture for classification of diabetic Retinopathy images *Comput. Electr. Eng.* **76** 56–64
[17] Adbillah B, Bustamam A and Sarwinda D 2017 Classification of diabetic retinopathy through texture features analysis *Int. Conf. Advanced Computer Science and Information Systems* (Piscataway, NJ: IEEE) pp 333–9
[18] Sarwinda D, Siswantining T and Bustamam A 2018 Classification of diabetic retinopathy stages using histogram of oriented gradients and shallow learning *2018 Int. Conf. on Computer, Control, Informatics and Its Applications (IC3INA)*
[19] Pratt H, Coenen F, Broadbent D M, Harding S P and Zheng Y 2016 Convolutional neural network for diabetic retinopathy *Procedia Comput. Sci.* **90** 200–5
[20] Wan S, Liang Y and Zhang Y 2018 Deep convolutional neural networks for diabetic retinopathy detection by image classification *Comput. Electr. Eng.* **72** 274–82
[21] Rahim S S, Palade V, Shuttleworth J and Jayne C 2016 Automatic screening and classification of diabetic retinopathy and maculopathy using fuzzy image processing *Brain Inform.* **3** 249–67
[22] Soebardi S *et al* 2011 Retina dari pediatrik hingga Geriatrik pada Bab: Retinopati diabetes perspektif endokrinolog *Jakarta Eye Center. Hal* 139–42
[23] Nayak J, Bhat P S, Acharya R U, Lim C M and Kagathi M 2008 Automated identification of diabetic retinopathy stages using digital fundus images *J. Med. Syst.* **32** 107–15
[24] Ting K M 2011 Confusion matrix *Encyclopedia of Machine Learning* ed K Ting M *et al* (Boston, MA: Springer) p 209

IOP Publishing

Photo Acoustic and Optical Coherence Tomography Imaging, Volume 2

Fundus imaging for the retina

Ayman El-Baz and Jasjit S Suri

Chapter 10

A framework for joint cup and disc segmentation in fundus images

Maleeha Khalid Khan and Syed Muhammad Anwar

10.1 Introduction

Automated medical image analysis began in the twentieth century, and it has revolutionized diagnosis and treatment. Medical images provide visual information on human body parts, and this information assists clinicians in decision-making. Medical images in the form of x-rays and 2D colored images have been used to diagnose diseases, and manual methods have been used to extract useful information from images. With advancements in technology, automatic detection and analysis techniques such as machine learning techniques have been developed. Medical images are being used in different machine learning techniques, with which the analysis of medical images can be carried out and useful information can be extracted [3]. Machine learning techniques extract important features from medical images and use these features to diagnose diseases and provide help in making clinical decisions [4]. Recently, many deep learning techniques have helped in the fast detection of abnormalities, classifying big data, and in segmentation. Deep learning techniques are now used to monitor chronic diseases using magnetic resonance imaging (MRI), in computed tomography (CT) for the detection of tumorous cells, and in ultrasound. For the detection of glaucoma, traditionally intraocular pressure and the cup-to-disc ratio of the retinal nerve are considered and manual assessments are done. The modern methods such as optical coherence tomography (OCT) can be used to map and measure the thickness of the retina and Heidelberg retinal tomography (HRT) can perform 3D analysis of the eye. Fundus images are being used by ophthalmologists and in advanced technologies to detect glaucoma, and these fundus images provide in-depth details about the structure of the eye, including blood vessels, retinal abnormalities in retinal screening, and optic nerve details. Deep learning

doi:10.1088/978-0-7503-2056-6ch10

systems can be used for analysis using fundus images and a model can be trained efficiently to detect glaucoma in its early stages.

Glaucoma is a cause of blindness and the damage caused by glaucoma, if prolonged and if not cured or properly treated, is permanent. Glaucoma is ranked as the second most common cause of blindness by the World Health Organization (WHO), the first being cataracts [1]. Glaucoma causes the optic nerve to deteriorate over time and if not cured in its early stages can later cause blindness. Glaucoma, according to its type, can be categorized into two classes: primary and secondary glaucoma. Both types of glaucoma have different symptoms and the damage they cause is different. There are 1.8 million people suffering from glaucoma in Pakistan (according to WHO reports, 2018) and among them are people who have lost their sight completely.

Ophthalmologists use the optic cup-to-disc ratio to diagnose glaucoma from fundus images. For automatic detection of glaucoma, the region of interest (ROI) is the central portion that defines the cup area that is considered. In glaucoma the optic cup-to-disc ratio is disturbed and the cup ratio increases. Recently, convolutional neural networks (CNNs) have been used for medical image segmentation for disease prediction, and hence automate manual segmetnaton to save time. CNN model visualization and CNN performance have been evaluated empirically and can be used to design high performance computer aided diagnostic (CAD) systems [7]. A deep learning model can detect and classify the ROIs and non-ROIs (NROIs). Deep learning models have been trained using fundus and gray scale images; the dataset of MESSIDOR has been used to classify ROI and non-ROI (nROI) images with an accuracy of 0.99 [2]. Ophthalmologists use manual assessment of the optic disc using ophthalmoscopy to manually examine the region of the optic disc. OCT is an imaging technique that is used by ophthalmologists for observing any disorder or abnormality in the optic nerve layer. OCT provides a detailed analysis of the eye, but these methods require an expert to make a diagnosis, making them time consuming. Structural changes in the optic nerve fiber layer can be observed by using fundus images, and deep learning techniques can automatically detect these structural changes in the images. An automatic technique is presented for segmenting the cup region to obtain the cup-to-disc ratio and to diagnose glaucoma, segmenting the disc and cup separately. The algorithm is a hybrid combination of different techniques [5] and is designed to first segment the disc, then suppress the blood vessels from fundus images, and then segment the cup region. The final analysis is performed by considering the cup-to-disc ratio.

For the cup-to-disc ratio (CDR), one the most used methods for the identification of glaucoma, a CDR $\geqslant 0.6$ is considered glaucomatous. As glaucoma deteriorates the optic nerve fiber and the damage is not curable in the later stages, earlier detection of glaucoma is necessary, and for this purpose the analysis of the optic disc, retinal layer, and optic nerve is necessary. The analysis information can help to detect the stage of glaucoma and help ophthalmologists make treatment decisions. The G-EYENET [5] multi-layered model is based on a deep convolutional autoencoder and CNN classifier. It first segments the colored fundus images using the autoencoder and it extracts the important features of glaucoma, and after segmentation the CNN classifier is used. The input images are first preprocessed using adaptive histogram equalization. U-Net is a specially designed architecture that is used in medical image analysis for segmentation. A modified U-Net architecture [6] can detect the center optic disc (OD) and make a probability map, and can detect the OD location. The modified U-Net uses a depth first search technique to achieve a higher probability map. A deep learning model that is capable of classifying glaucoma [7] and can detect optic disc abnormalities using CNN, our model uses a cascade classifier, the AdaBoost algorithm, for feature extraction and training classifiers, and is trained on the DRIVE, STARE, DIARETDB1, and MESSIDOR databases.

In glaucoma the optic nerve is damaged which can be visually seen from fundus images. For earlier detection of glaucoma, a hybrid neural network-based algorithm that is based on adaptive histogram equalization is used. In a hybrid color and structure descriptor (HCSD) [8] model, the model used adaptive histogram equalization and a hybrid combination of feature vectors. The HCSD networks work by extracting significant features for the detection of glaucoma and these features are later used for the classification of glaucoma. Color fundus images are commonly used in the detection of glaucoma, as the area of the cup and disc are discriminated by color. A U-Net [9] based fully connected convolutional neural network was trained on MESSIDOR and then shape features were extracted. Images were first preprocessed then segmentation was performed using the U-Net model. False positives were removed from the segmented images using morphological opening. The morphological opening used erosion to remove foreground pixels from edges, and the model was cross-validated on the MESSIDOR dataset. High intraocular pressure (IOP) damages the retinal rim and this can directly damage vision; this damage if not detected in the early stages can even cause blindness. 2D and 3D images can be used for the assessment of the retinal layer, for early detection of glaucoma. Among the most common methods for the detection of glaucoma is the segmentation of the optic disc and optic cup using colored fundus images. The optic disc and optic cup-to-disc ratio are not the only parameters in detecting glaucoma, the neuroretinal rim thickness is also a major factor. Hence segmenting and analyzing blood vessels in the eye is also significant along with identifying the OD and OC ratio. U-Net modified and preprocessed images were given as input for segmentation and localization [10] based on structural and gray level properties, using fundus images of the open source dataset RIMONE. Most segmentation methods have used the optic disc and optic cup ratio separately, but OD and OC joint segmentation can improve segmentation methods [11]. The M-Net architecture, which consists of multiple input layers and a U-shaped convolution layer, has a rich hierarchical structure used for earlier classification. M-Net is used for segmentation; it consists of U-Net with an end-to-end learner added. M-Net can help to solve OD and OC segmentation jointly. The M-Net architecture has been trained on the ORIGA and SCES datasets, which are not open source datasets. Joint M-Net with polar transformation has better accuracy compared to a simple U-Net model. The optic nerve head (ONH) is the area of the eye that has all the connecting tissues in it; a slight change in these tissues is an indicator of glaucoma. Images show the structural change in the ONH tissues, and both the retinal layer and optic nerve change can be observed more clearly using 3D images. OCT is a technique to visualize the eye, and OCT images are used to analyze the tissues in the eye. The change in tissue texture compared to normal texture and the change in their spatial arrangement show the presence of glaucoma [12]. U-Net is a widely used network in biomedical image segmentation. An extension of U-Net, bidirectional Conv-LSTM (long short term memory) [13], uses dense convolutions instead of simple convolutions to make the network learn more features. Bidirectional Conv-LSTM, between skip connections, considers spatial correlation, and discriminative information improving the segmentation results. DRUNET [12] is a modified U-Net, more focused on contextual features and the connective tissues of ONH. 3D images provide more detail for medical analysis and complete details of human organs. The latest methods of biomedical analysis use 3D technology. Using a modified [14] U-Net that has been trained on 3D images, U-Net 3D volumetric segmentation can be performed. The modified U-Net architecture was trained to segment 3D images, with an architecture based on volumetric segmentation and trained on open source images.

Diabetic retinopathy (DR) is a form of glaucoma and occurs in diabetic patients; its adverse effects are difficult to control. CNN is used to diagnose DR using fundus images. DR can also be detected using color fundus images, as in [15]. The CNN was trained on the Kaggle dataset and a simple layered model was used. Due to a lack of sufficient data, the pixel method was used. Images were first fed into the support vector machine (SVM) classifier to extract variations of shapes. Most techniques are focused on separate cup and disc segmentation, but joint optic cup and disc segmentation helps in retaining features which are lost in separate cup and disc segmentation. The deep learning model generative adversarial networks (GANs) [16] uses a discriminator as a verifier of real and fake segmentation. The two main functions of GANs are the generator and discriminator; the generator makes a segmentation map and the discriminator generates a ground truth segmentation map used to evaluate the segmentation map. GANs achieve more accuracy than the standard methods of segmentation.

For supporting local hospitals without advanced technologies, a semi-automated system for the detection of glaucoma can be achieved using a gradient vector flow (GVF) based snake active contour method [24]. GVF considers the size and shape of the optic disc and cup and an ellipse fit is used to find the CDR, considering the inferior, superior, nasal, and temporal rims (the ISNT rule). The GVF snake method identifies the area of the cup and disc, and a further ellipse fit is applied to accurately determine the cup and disc area, to calculate the CDR. A CNN is based on the available data and labeled data so the model can learn features and make predictions. Transfer learning is CNN based [25] and it uses 3D images and can detect variation in retinal nerve fiber layer thickness. Transfer learning uses the concept of self-learning. Transfer learning can provide better and more effective analysis in detecting glaucoma. VGG-16 is a CNN architecture used for object classification and object detection. The combined effect of VGG-16s based on the disc center [26], retinal nerve fiber layer (RNFL) thickness, ganglion cell complex (GCC) thickness, RNFL deviation map, and GCC deviation map is used in transfer learning, this model covers different techniques to give better performance. Semi-supervised transfer learning is an approach that uses the CNN based ImageNet model [27], which is pre-trained model, for glaucoma detection then uses these results for self-training. Self-training is a concept of transfer learning, involving understanding and predicting labels from images, and self-learning classifiers can be used for unlabeled data. Semi-supervised learning methods achieve more accuracy than supervised methods of learning, and this is evaluated using training sets by changing the layers of the CNN model and adjusting them in a way to make a model that can learn features from the human eye and is able to make a prediction about the presence of glaucoma. For evaluating the quality of optic disc and optic cup segmentation, the self-assessment method uses an automated diagnosis system [17]. Based on super pixel classification, the method in proposed [18] was used for glaucoma screening and the CDR was used for the detection of glaucoma; the F-score was improved in this automated system. The CDR is one of the best methods used for the analysis and detection of glaucoma. The ISNT rule [19] states that the neuroretinal rim, which is in order inferior ⩾ superior ⩾ nasal ⩾ temporal, of a glaucomatous affected eye violates this rule. The areas of the neuroretinal rim (NRR) are also used for verification of glaucoma, and a cup detection hill climbing algorithm [20] is used and for cup extraction fuzzy C-mean clustering is used. The combined effect of these techniques can give better accuracy compared to manual methods of detection of glaucoma. In the neuro-fuzzy inference system (ANFIS) technique [21] to train an artificial neural network, twenty-one parameters were used to classify different types of glaucoma. The ANFIS based system is more dependent on larger data for better results, so it is less efficient than a deep learning system. Structural changes such as changes in retinal layer thickness and changes in the optic cup are also key contributors in detecting glaucoma. Structural changes of the eye appear earlier compared to functional features, which appear at later stages [22]. A novel algorithm based on optic disc localization and segmentation based on structural feature detection using a hybrid feature extractor and SVM to classify glaucoma is presented in [22]. Glaucoma can also damage the blood vessels in the eyes. Optic disc localization and vessel-based segmentation [23] along with an SVM classifier can help in detecting retinal abnormalities

The CNN based model [28] is trained on images that are first converted from color images to gray scale images as the information contained in gray scale images is enough for classification. The CNN based model then uses a classifier to learn features and classify them. A fully connected seven layered CNN based model is trained for detecting the joint optic disc and fovea locations. The root mean square error (RMSE) is used to evaluate CNN based model performance. Fundus images provide in-depth details about the eye, and the disc, cup, and vessels can be observed using fundus images. Separating the optic cup and disc, and vessel detection for glaucoma has been done using disc segmentation with a threshold along with a histogram [29]. The threshold improved segmentation results for the disc and cup segmentation, done using ISNT division (used by ophthalmologists), but the method failed to segment vessels. The deep learning-based technique automatic feature learning for glaucoma detection based on deep learning (ALADDIN) [28] was used to capture distinguishing features to make the model learn the pattern of glaucoma. Two linear CNN models were used; the output of the first CNN was the input to the second, the CNN classifier. The two models were used to learn hidden patterns and were based on a textualizing training strategy. The ALADDIN model uses the two CNNs, the first is able to learn features trained using backpropagation. The second CNN used in the ALADDIN model is a multilayer perceptron, a deep model itself, based on feature re-use. The ALADDIN model was trained on the ORIGA and SCES datasets. Computer aided diagnostics (CADx) [30] uses a contextual training strategy through which the model can learn a deep pattern method based on the unsupervised CNN based model for segmenting the optic disc. The segmented output from the CNN is then used by a deep-belief network (DBN) model for feature learning and the softmax classifier is used to classify glaucoma. A MATLAB [29] based system can be used to detect glaucoma but it requires a combination of techniques for proper segmentation of glaucoma. Fundus images are first preprocessed and converted into gray scale images, then the optic disc and optic cup are regions are detected separately.

Ellipse fitting is used for detecting optic disc boundaries, then the radius area of the disc is detected to calculate the CDR [31]. A MATLAB based method requires a lot of preprocessing, making it less efficient than a deep learning-based system. A modified U-Net model [31] for segmenting the OD and OC with high precision uses spline interpolation and histogram equalization techniques. The modified U-Net uses preprocessed images allowing better training of the model and increasing the performance of the system. The CNN based VGG-19 model [32] was used to classify ten different types of glaucoma. The CNN based VGG-19 model was used with MatConvNet (a MATLAB toolbox implementation) for automatic detection of disturbances in the retina. A normal layer based CNN classifier is not able to give high accuracy for detecting different types of glaucoma. For this, transfer learning incorporated with a CNN classifier was used, but the study failed to give better results due to the lack of dataset availability for different types of glaucoma [33]. The VGG-16, ImageNet, ResNet, and Inspection models are pre-trained models used for classification purposes, and these models are being used in applications. CNN based deep learning models are used to classify glaucoma and the random forest classifier [34] is also used with each model to improve accuracy and over-fitting. The performance of each model is evaluated to determine which model is best suited to classifying glaucoma. VGG-16 with the random forest classifier is proposed as the best model for classification.

The front part of the eye contains the cornea and lens, and the space between them is filled with a fluid called aqueous humor. New aqueous humor is constantly being produced and then drained out of the eye. To maintain proper pressure inside the eye the amount produced and the amount drained must be kept in balance. Glaucoma involves damage to the structures that allow the aqueous humor to drain out. There are two outlets for the aqueous humor to drain: the trabecular meshwork and the uveoscleral outflow. In open-angle glaucoma, the trabecular meshwork offers increased resistance to fluid outflow [35]. This causes pressure to build up inside the eye. In closed-angle glaucoma, both the uveoscleral drain and the trabecular meshwork are blocked. For detecting open- and closed-angle glaucoma machine learning approaches can be used. In [36] a review of different machine learning techniques is presented including fuzzy logic, neural networks, *k*-nearest neighbors, canny edge detector, and SVM. Fuzzy logic and linear regression can predict glaucoma and CNN is the best at classifying glaucoma. When detecting glaucoma in images, the efficiency of the system is reduced due to redundant data. A CNN based AG-CNN (attention-based) classifier for glaucoma detection [37] was trained on a large dataset of more than eleven thousand fundus images. Supervised learning is used and the AG-CNN architecture can locate pathological areas on fundus images. The authors of [37] have proposed attention maps that are made using an attention prediction subnet, this proposed model works best in detecting glaucoma.

For optic disc segmentation, a CNN based on P-Net [38] is used. P-Net is used to generate a more accurate optic disc segmentation map. P-Net is used to generate a low-resolution segmentation map that is used to train Fine-Net. P-Net is trained on the Drishti, MESSIDOR, and DRINOS databases. P-Net gives a high Dice coefficient. The retinal nerve fiber layer defect (RNFLD) is also an early indicator of glaucoma. Fundus images can be used for RNFL defect detection. RNFL is visible as it appears to be radiating from the optic disc region. Proposed by the authors of [39], an experimental study based on boundary localization was used for RNFL detection. The boundary identifies the area of interest and this can be done using the Laplace, Sobel, and Canny methods. Examining these three boundary detection techniques most commonly used for the detection of the cup and disc area in fundus images, a comparative study conducted in [40] shows that the Laplace method has flexibility but it adds noise in the resultant images, Sobel is speedy method but it has less sensitivity, and Canny is a good localization method but it is complex for use for segmentation purpose. An automatic system is used to diagnose glaucoma, using a twin support vector machine (TWSVM) classifier [41]. The TWSVM is compared with normal CNN classifiers for classifying glaucoma. The TWSVM classifier is a modified form of the SVM classifier, which builds a positive hyperplane and a negative hyperplane, which is used in decision making in predicting glaucoma and non-glaucoma. From fundus images, the region of interest (ROI) is considered the area in which the cup and disc are located, the optic disc region. One technique detects the ROI by applying separate red and green channels on ROI images, then pre-processes them for segmentation [42], and can locate an area of the red channel and red disc similar to an area of the green channel and green disc. The CDR ratio can be best calculated using this technique, as this proposed technique is simple and can better discriminate between glaucoma and non-glaucoma images.

The ISNT rule is directly related to neuroretinal rim thickness. An automatic technique [43] for processing fundus images for the detection of glaucoma works by preprocessing the images, then the ISNT rule is applied, and then localization of the optic cup and optic disc is performed. For accurate localization of OD, a square template range is applied to detect the affected regions to detect glaucoma. Accurate localization is achieved using the square template region, which ignores the drusen region. The algorithm requires more processing time compared to the CNN. For optic disc segmentation, *K*-means clustering is proposed to find irregular objects, and clustering is done based on colored objects [44]. Edge detection is performed to reduce the extra data in the image while preserving the image structure [45]. The proposed edge detection algorithm achieves a boundary detection accuracy of 92% of disc and cup region.

10.2 Glaucoma detection using REFUGE data

The dataset used in this research are from the REFUGE and REFUGE2 challenge (available at https://refuge.grand-challenge.org/); the data are not open source. The data are obtained by participating in the REFUGE challenge. REFUGE2 data are used for disc and cup segmentation. Along with this the dataset contains test images of glaucoma and non-glaucoma so the calculated CDR can be cross-validated. REFUGE contains a total of 1200 images—400 training, 400 testing, and 400 validation images. The training images contain 40 glaucoma and 360 non-glaucoma images. The high imbalance of glaucoma and non-glaucoma images makes classification challenging. In preprocessing the images are converted into polar coordinates and for each image five images are generated in polar coordinates making two thousand images. REFUGE2 contains the ground truth for validation and testing images as well. The data lack a ground truth CDR ratio. The data can be cross-validated by using ground truth cup and disc mask images. For validation of the segmentation results the Dice coefficient is used in this research. The Dice coefficient depends on the overlapping area divided by the total number of pixels in both images.

10.2.1 Preprocessing REFUGE2 data

Fundus images from REFUGE data are first preprocessed and converted into polar coordinates for each fundus image, and a polar transformed image is generated. The image ground truth is also generated in polar coordinates based on disc and cup mask images. The polar transform gives a pixel-wise converted image. The fundus image is not directly used for the segmentation process, as the region of interest needs to be identified. Based on a literature study the polar transformed image is the best preprocessing method. The polar transformation makes the fundus image flat based on the optic disc center and increases the ratio of the cup region making the image more balanced compared to the original fundus image. In normal fundus images the optic cup appears inside the disc region whereas in the polar transformed image the cup and disc region appear in a layer-like structure. The cup, disc, and background in the polar transformed image appear in an ordered layer-like structure and this balanced polar transformed image helps in model training by avoiding over-fitting and improves segmentation results. The polar transformation obtains spatial constraints and helps in better optic disc and cup segmentation (figures 10.1 and 10.2).

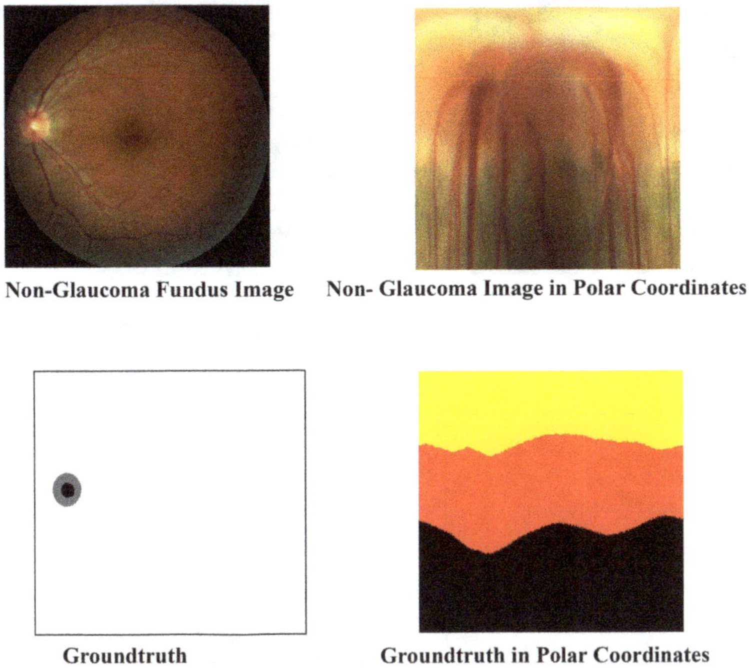

Figure 10.1. Fundus image of non-glaucoma and its conversion into polar coordinates and ground truth in a polar transform.

10.2.2 Relation between fundus image and polar transformed image

Polar transformation is used to convert the image into a pixel-wise representation so better analysis can be performed. As shown in figure 10.3 $L(u, v)$ denotes the point on the fundus image plane where the origin is set as the disc center $O(u_o, v_o)$, and (u, v) are the Cartesian coordinates. The corresponding point on the polar coordinate system is $L'(\theta, r)$, where r denotes radius and θ denotes the directional angle of the original point p. The transform relation between the polar and Cartesian coordinates is:

$u = r \cos \theta$ where, $r = \sqrt{u^2 + v^2}$,

$v = r \sin \theta$.

The height of the transformed images is R and the width is 2π (figure 10.3).

10.2.3 Implementation

Our proposed model is implemented based on Python with Keras and also with a Tensorflow backend, and it uses MATLAB to calculate the CDR ratio. The results are processed on Linux 18.04.2 lTS (GNU/Linux 4.15.07 generic x86_64).

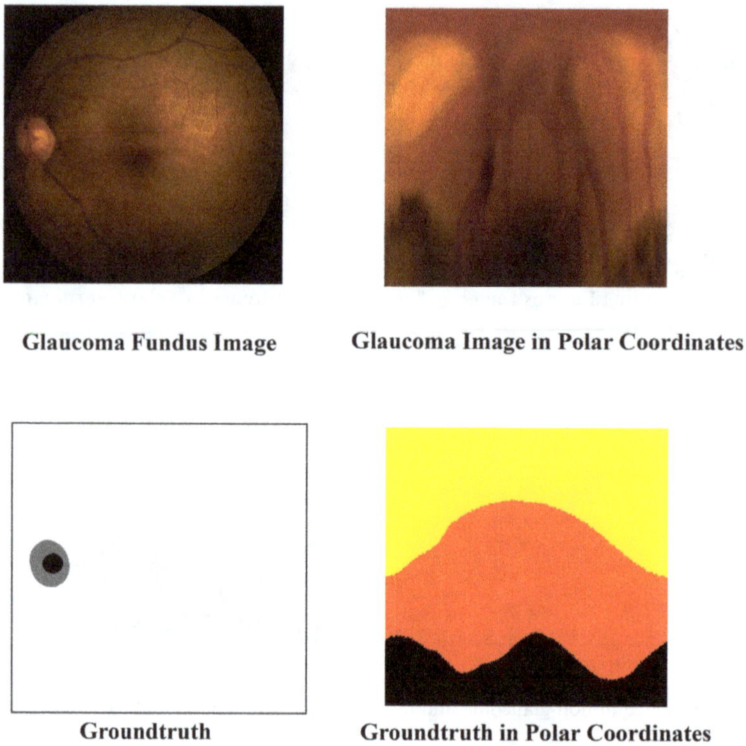

Figure 10.2. Fundus image of glaucoma and its conversion into polar coordinates and ground truth in polar coordinates.

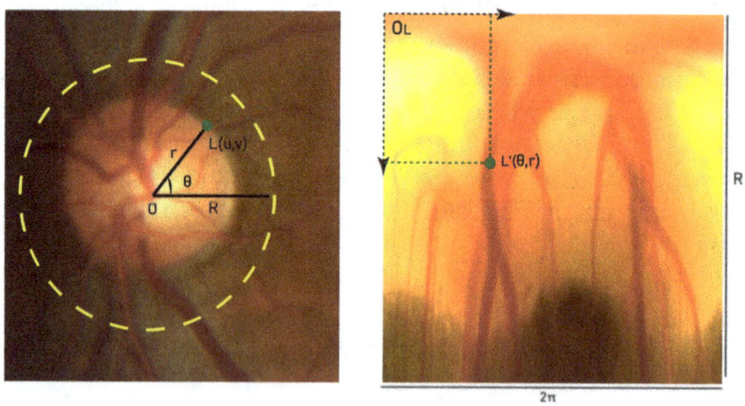

Figure 10.3. Relationship between the spatial image and polar transformed image.

10.2.3.1 Segmentation experiments

In image segmentation the image is converted into segments, which represent an object or part of an object. Segmentation is done to classify the different objects in an image and to detect an object or its parts in an image. Medical image segmentation is focused on finding regions of interest based on specific descriptions. The datasets used for this experiment are REFUGE and REFUGE2. REFUGE contains 400 images with their ground truths, out of these 40 are glaucoma images and 360 are non-glaucoma images making proportions of 10% glaucoma images and 90% non-glaucoma images. REFUGE2 contains 1200 images with their ground truths, these images are also unbalanced test images and are also made up of 10% glaucoma images and 90% non-glaucoma images. These REFUGE2 data are subjected to segmentation after preprocessing. We have used data of 800 images. In preprocessing the images are converted into polar coordinates and their ground truth, that is cup and disc mask images, are also converted into polar coordinates. After being preprocessed the images are given to the model for training. The model has convolutional layers and the encoder path has convolutional layers that extract features. The encoder path has pooling layers to reduce the spatial size of the image, ultimately reducing the number of parameters and the amount of computation in the network. The skip connections transfer the feature maps from encoder to decoder, and between the skip connections a bidirectional convolutional LSTM is used. The bidirectional LSTM trains two instead of one LSTM on input sequences, one the input sequence and second a reversed copy of the input sequence.

The bidirectional Conv-LSTM considers the spatial correlation, which is ignored in a simple LSTM. The bidirectional Conv-LSTM processes the input data in the forward and backward directions. Both forward and backward dependencies can improve the network performance [13]. The improved results of segmentation in this research work shows that adding the bidirectional Conv-LSTM between skip connections captures more discriminative features and enhances the result of the segmentation process. On the decoder side the outputs from all the decoder layers have been taken and an average of all the layers is taken.

10.2.3.2 Post-processing of data

After segmentation the task was to convert the images back into spatial coordinates. For this post-processing is performed using segmentation trained results. This process requires extracting the cup and disc, from which post-processing mask images were generated. Post-processing is performed by loading the images and then detecting the disc region using U-Net, using disc and cup segmentation from the information of the trained model MBLSTM, and finally extracting mask images. Masks were extracted using the inverse of the polar transformation which considers the disc map and cup map.

The black area shows the background, the gray region is the optic disc, and inside the optic disc is the white part, which is the optic cup that needs to be further processed to obtain the correct optic disc and cup region (figure 10.4). Some of the mask images were disturbed, and the areas of the cup and disc were not clear. In a few images the cup area was disturbed and in a few images the disc area was disturbed (figure 10.5).

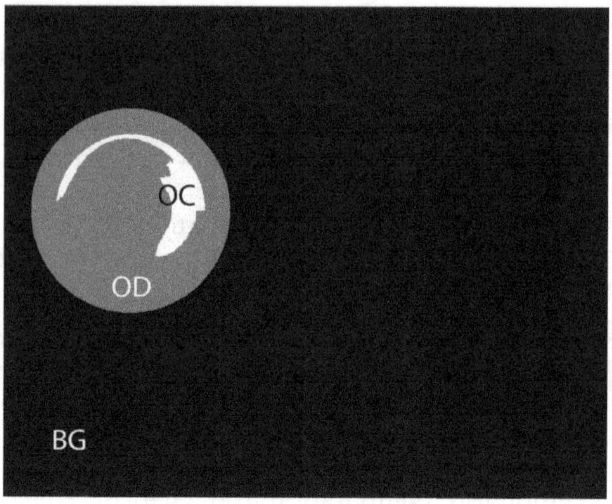

Figure 10.4. Generated mask image.

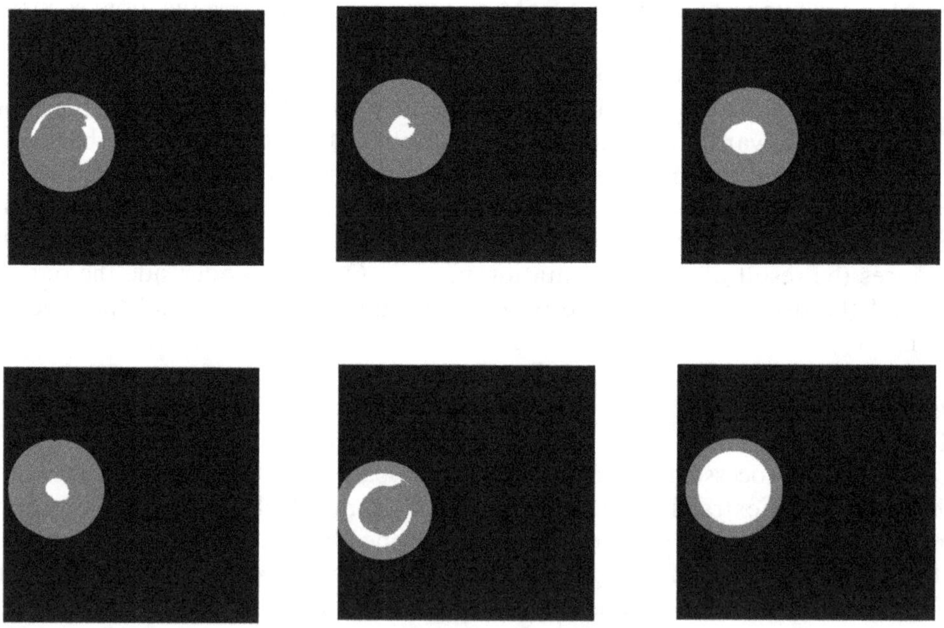

Figure 10.5. Disturbed mask images.

10.2.3.3 MATLAB post-processing

The mask images were then subjected to further post-processing using MATLAB where images were reconstructed. Using the MATLAB toolbox, the cup and disc area were identified. The images after post-processing were in bmp format and some images were disturbed and could not be reconstructed as the cup area was not

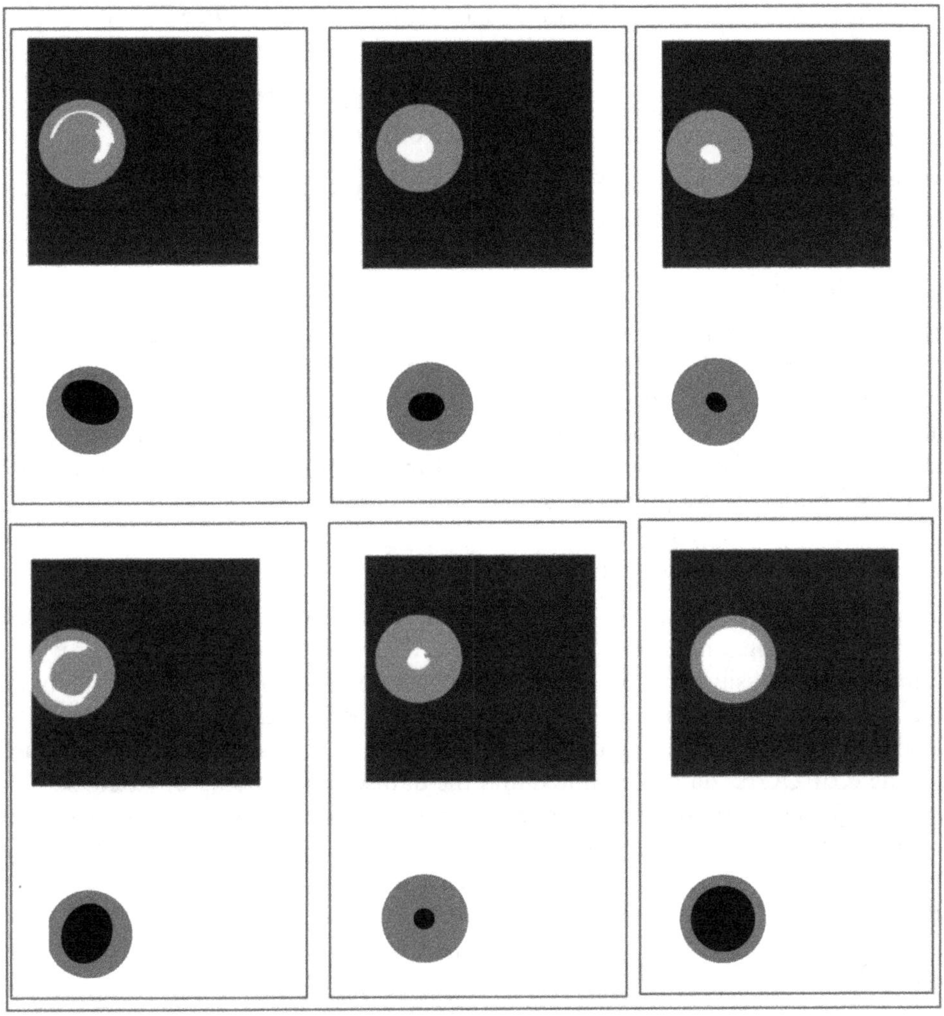

Figure 10.6. Reconstructed mask images.

identifiable. About 5% images were disturbed and could not be reconstructed. The rest of the images were reconstructed. Some reconstructed images are shown in figure 10.6.

There were a few cases in which the disturbed mask images could not be reconstructed as the area of cup was not defined, these images were removed.

10.2.3.4 Dice coefficient

All the resultant images obtained after segmentation and post-processing were used to calculate the Dice coefficient. The segmented cup disc mask images were compared with the ground truth images, and the Dice coefficient was calculated using the overlapping area between the cup disc mask images and the ground truth images. The standard formula for the Dice coefficient was used,

$$\text{Dice coefficient} = \frac{2 \times \text{Overlapping area}}{\text{Total number of pixels in both images}}.$$

The proposed MBLSTM achieves an optic disc average Dice coefficient value of 0.93 and value of the optic cup Dice coefficient is 0.88.

10.3 Conclusion

In this research a modified U-Net is proposed that is based on joint segmentation of the disc and cup with bidirectional convolution BLSTM, to combine feature maps from the encoding and decoding ends, and based on the segmented output abnormalities in the cup-to-disc ratio can be observed. The modified U-Net gives segmented cup-to-disc images. The proposed model MBLSTM has been tested on REFUGE2 data; the proposed model achieves an accuracy of 98%.

10.3.1 Limitations of the research

Our approach for segmentation of the cup and disc from fundus images requires a lot of preprocessing, and after segmentation is done a lot of post-processing work is required. Preprocessing and post-processing limit research to only specific data and make segmentation limited to the cup and disc. REFUGE2 was the only data available for cup and disc segmentation and other datasets were not open source. Also, the REFUGE2 data are limited and the dataset is still being updated, so it has more non-glaucoma fundus image data compared to glaucoma data. This affects the segmentation as well as classification results. Further, REFUGE2 data do not provide a ground truth CDR making it further difficult to cross validate the CDR results with standards. The CDR calculated for the REFUGE2 ground truth does not indicate classification clearly, making this research limited to only segmentation and detection.

References

[1] Thylefors B and Negrel A D 1994 The global impact of glaucoma *Bull. World Health Organiz.* **72** 323

[2] Chai Y, Liu H and Xu J 2018 Glaucoma diagnosis based on both hidden features and domain knowledge through deep learning models *Knowl.-Based Syst.* **161** 147–56

[3] Kim J, Candemir S, Chew E Y and Thoma G R 2018 Region of interest detection in fundus images using deep learning and blood vessel information *IEEE 31st Int. Symp. on Computer-Based Medical Systems*

[4] Dutta K, Mukherjee R, Kundu S, Biswas T and Sen A 2018 Automatic evaluation and predictive analysis of optic nerve head for the detection of glaucoma *2nd Int. Conf. on Electronics, Materials Engineering and Nano-Technology*

[5] Pal A, Moorthy M R and Shahina A 2018 G-EYENET: a convolutional autoencoding classifier framework for the detection of glaucoma from retinal fundus images *25th IEEE Int. Conf. on Image Processing*

[6] Gu Z, Jiang S, Lee J, Xie J, Cheng J and Liu J 2018 Automatic localization of optic disc using modified U-Net *Proc. 2018 Int. Conf. on Control and Computer Vision*

[7] Alghamdi H S, Tang H L, Waheeb S A and Peto T 2016 Automatic optic disc abnormality detection in fundus images: a deep learning approach *Ophthalmic Medical Image Analysis International Workshop* vol 3, No. 2016) (Iowa City: University of Iowa)

[8] David D S and Jayachandran A 2018 A new expert system based on hybrid colour and structure descriptor and machine learning algorithms for early glaucoma diagnosis *Multimedia Tools Appl.* **79** 5213–24

[9] Sadhukhan S, Ghorai G K, Maiti S, Karale V A, Sarkar G and Dhara A K 2018 Optic disc segmentation in retinal fundus images using fully convolutional network and removal of false-positives based on shape features *Deep Learning in Medical Image Analysis and Multimodal Learning for Clinical Decision Support* (Berlin: Springer) pp 369–76

[10] Mittapalli P S and Kande G B 2016 Segmentation of optic disk and optic cup from digital fundus images for the assessment of glaucoma *Biomed. Signal Process. Control* **24** 34–46

[11] Fu H, Cheng J, Xu Y, Wong D W K, Liu J and Cao X 2018 Joint optic disc and cup segmentation based on multi-label deep network and polar transformation *IEEE Trans. Med. Imaging* **37** 1597–605

[12] Devalla S K *et al* 2018 DRUNET: a dilated-residual U-Net deep learning network to segment optic nerve head tissues in optical coherence tomography images *Biomed. Opt. Express* **9** 3244–65

[13] Azad R, Asadi-Aghbolaghi M, Fathy M and Escalera S 2019 Bi-directional ConvLSTM U-Net with Densley connected convolutions *Proc. IEEE Int. Conf. on Computer Vision Workshops*

[14] Çiçek Ö, Abdulkadir A, Lienkamp S S, Brox T and Ronneberger O 2016 3D U-Net: learning dense volumetric segmentation from sparse annotation *Int. Conf. on Medical Image Computing and Computer-assisted Intervention*

[15] Pratt H, Coenen F, Broadbent D M, Harding S P and Zheng Y 2016 Convolutional neural networks for diabetic retinopathy *Proc. Comput. Sci.* **90** 200–5

[16] Shankaranarayana S M, Ram K, Mitra K and Sivaprakasam M 2017 Joint optic disc and cup segmentation using fully convolutional and adversarial networks *Fetal, Infant and Ophthalmic Medical Image Analysis* (Berlin: Springer) pp 168–76

[17] Atheesan S and Yashothara S 2016 Automatic glaucoma detection by using funduscopic images *Int. Conf. on Wireless Communications, Signal Processing and Networking*

[18] Cheng J, Liu J, Xu Y, Yin F, Wong D W K, Tan N-M, Tao D, Cheng C-Y, Aung T and Wong T Y 2013 Superpixel classification based optic disc and optic cup segmentation for glaucoma screening *IEEE Trans. Med. Imaging* **32** 1019–32

[19] Harizman N, Oliveira C, Chiang A, Tello C, Marmor M, Ritch R and Liebmann J M 2006 The ISNT rule and differentiation of normal from glaucomatous eyes *Arch. Ophthalmol.* **124** 1579–83

[20] Ahmad H, Yamin A, Shakeel A, Gillani S O and Ansari U 2014 Detection of glaucoma using retinal fundus images *Int. Conf. on Robotics and Emerging Allied Technologies in Engineering*

[21] Diptu N A *et al* 2018 Early detection of glaucoma using fuzzy logic in Bangladesh context *Int. Conf. on Intelligent Systems*

[22] Salam A A, Akram M U, Wazir K and Anwar S M 2015 A review analysis on early glaucoma detection using structural features *IEEE Int. Conf. on Imaging Systems and Techniques*

[23] Salam A A, Akram M U, Abbas S and Anwar S M 2015 Optic disc localization using local vessel based features and support vector machine *IEEE 15th Int. Conf. on Bioinformatics and Bioengineering*

[24] Kusumandari D E, Redhyka G G and others 2015 The comparison of GVF snake active contour method and ellipse fit in optic disc detection for glaucoma diagnosis *Int. Conf. on Automation, Cognitive Science, Optics, Micro Electro-Mechanical System, and Information Technology*

[25] Al-Bander B, Al-Nuaimy W, Al-Taee M A and Zheng Y 2017 Automated glaucoma diagnosis using deep learning approach *14th Int. Multi-Conf. on Systems, Signals and Devices*

[26] An G, Omodaka K, Hashimoto K, Tsuda S, Shiga Y, Takada N, Kikawa T, Yokota H, Akiba M and Nakazawa T 2019 Glaucoma diagnosis with machine learning based on optical coherence tomography and color fundus images *J. Healthc. Eng.* **2019** 4061313

[27] Al Ghamdi M, Li M, Abdel-Mottaleb M and Abou Shousha M 2019 Semi-supervised transfer learning for convolutional neural networks for glaucoma detection *IEEE Int. Conf. on Acoustics, Speech and Signal Processing*

[28] Al-Bander B, Al-Nuaimy W, Al-Taee M A, Al-Ataby A and Zheng Y 2016 Automatic feature learning method for detection of retinal landmarks *9th Int. Conf. on Developments in eSystems Engineering*

[29] Nikam S M and Patil C Y 2017 Glaucoma detection from fundus images using Matlab GUI *3rd Int. Conf. on Advances in Computing, Communication and Automation*

[30] Abbas Q 2017 Glaucoma-deep: detection of glaucoma eye disease on retinal fundus images using deep learning *Int. J. Adv. Comput. Sci. Appl.* **8** 41–5

[31] Joshua A O, Nelwamondo F V and Mabuza-Hocquet G 2019 Segmentation of optic cup and disc for diagnosis of glaucoma on retinal fundus images *Southern African Universities Power Engineering Conf./Robotics and Mechatronics/Pattern Recognition Association of South Africa*

[32] Choi J Y, Yoo T K, Seo J G, Kwak J, Um T T and Rim T H 2017 Multi-categorical deep learning neural network to classify retinal images: a pilot study employing small database *PLoS One* **12** e0187336

[33] Ahn J M, Kim S, Ahn K-S, Cho S-H, Lee K B and Kim U S 2018 A deep learning model for the detection of both advanced and early glaucoma using fundus photography *PLoS One* **13** e0207982

[34] Thakoor K A, Li X, Tsamis E, Sajda P and Hood D C 2019 Enhancing the accuracy of glaucoma detection from OCT probability maps using convolutional neural networks *41st Annual Int. Conf. of the IEEE Engineering in Medicine and Biology Society*

[35] Weinreb R N and Khaw P T 2004 Primary open-angle glaucoma *Lancet* **363** 1711–20

[36] Khalil T, Khalid S and Syed A M 2014 Review of machine learning techniques for glaucoma detection and prediction *Science and Information Conf.*

[37] Li L, Xu M, Liu H, Li Y, Wang X, Jiang L, Wang Z, Fan X and Wang N 2019 A large-scale database and a CNN model for attention-based glaucoma detection *IEEE Trans. Med. Imaging* **39** 413–24

[38] Mohan D, Kumar J H and Seelamantula C S 2019 Optic disc segmentation using cascaded multiresolution convolutional neural networks *IEEE Int. Conf. on Image Processing*

[39] Panda R, Puhan N B, Rao A, Padhy D and Panda G 2017 Recurrent neural network based retinal nerve fiber layer defect detection in early glaucoma *IEEE 14th Int. Symp. on Biomedical Imaging*

[40] Soltani A, Battikh T, Jabri I, Mlouhi Y and Lakhoua M N 2016 Study of contour detection methods as applied on optic nerve's images for glaucoma diagnosis *Int. Conf. on Control, Decision and Information Technologies*

[41] Touahri R, Azizi N, Benzebouchi N E, Hammami N E and Moumene O 2018 A comparative study of convolutional neural network and twin SVM for automatic glaucoma diagnosis *Int. Conf. on Signal, Image, Vision and their Applications*

[42] Patel S C and Patel M I 2018 Analysis of CDR of fundus images for glaucoma detection *2nd Int. Conf. on Trends in Electronics and Informatics*

[43] Pinos-Velez E, Flores-Rivera M, Ipanque-Alama W, Herrera-Alvarez D, Chacon C and Serpa-Andrade L 2018 Implementation of support tools for the presumptive diagnosis of glaucoma through identification and processing of medical images of the human eye *IEEE Int. Systems Engineering Symp.*

[44] Poshtyar A, Shanbehzadeh J and Ahmadieh H 2013 Automatic measurement of cup to disc ratio for diagnosis of glaucoma on retinal fundus images *6th Int. Conf. on Biomedical Engineering and Informatics*

[45] Shetty S C and Gutte P 2018 A novel approach for glaucoma detection using fractal analysis *Int. Conf. on Wireless Communications, Signal Processing and Networking*